Ben Stacy Jerrik (Ed.)

Nitin Pradhan

Ben Stacy Jerrik (Ed.)

Nitin Pradhan

Chief information officer, United States Department of Transportation, Presidency of Barack Obama

Part Press

Publisher:
Part Press is a trademark of
International Book Market Service Ltd., 17 Rue Meldrum, Beau Bassin, 1713-01 Mauritius
Email: info@bookmarketservice.com
Website: www.bookmarketservice.com

Published in 2012

Printed in: U.S.A., U.K., Germany. This book was not produced in Mauritius.

ISBN: 978-620-1-61990-6

Contents

Articles

References

Nitin_Pradhan

Nitin Pradhan		
[[file:Nitin Pradhan, Chief Information Officer.jpg		alt=]]
Departmental Chief Information Officer (CIO)		
Incumbent		
Assumed office July 6, 2009		
Preceded by	Dan Mintz	
Prior position: IT executive, Fairfax County Public Schools (FCPS)		
Personal details		
Website	US DOT CIO [1]	

Nitin Pradhan (Hindi: नतिनि पर्‌धान;) was sworn in on July 6, 2009, as the Departmental Chief Information Officer (CIO) for the US Department of Transportation (DOT) as part of the Obama Administration.

Pradhan is the chief advisor to Secretary of Transportation, Ray LaHood relating to information technology. In his role as the Departmental CIO, Pradhan provides IT vision, strategy, planning, policy and oversight for DOT's more than $3.0 billion IT portfolio,[2] the 6th largest in the Federal Government. DOT has over 55,000 employees and $70 billion in its budget. Prior to joining DOT, Pradhan was an IT Executive at Fairfax County Public Schools (FCPS), the 12th largest school district in the USA.

Earlier, Pradhan was the Managing Director of Virginia's Center for Innovative Technology (CIT).[3] He is also the co-founder and former CEO of a wireless startup. Pradhan's personal blog is titled "Winning Through Technology."[4]

Early life and education

Born and brought up in Pune, India, Pradhan attended Loyola High School (Pune), where he was known as an avid sportsman focused on athletics, including soccer and field hockey. After high school, Pradhan attended the Faculty of Technology and Engineering, Maharaja Sayajirao University of Baroda, India where he completed his bachelor's degree in engineering. He followed this degree with master's in marketing management from University of Pune, India.

After receiving his masters, Pradhan started his technology career in India by establishing and growing one of the most profitable divisions of International Data Management Ltd., (IDM) in India. Pradhan came to Washington, DC on a graduate fellowship from the Kogod School of Business at The American University (AU), to study for his second master's degree, accounting.

IT philosophy

Pradhan's IT philosophy focuses on people first, innovation, agility and driving business value. He believes that technology is first about people. Technology based Innovation is his second theme. He believes that Chief Information Officers (CIOs) have a dual role to play as Chief Innovation Officers. This aspect focuses on creating an innovation life cycle within organizations consisting of ideation and crowd sourcing tools like DOT's IdeaHub,[5] a process for selection of best ideas, and a cloud based agility platform for quick deployment of "apps" that drive business value. Finally, he has promoted the concept of new IT—IT 2.0,[6] as he calls it, is based on immediately providing significant business value.

He is a strong proponent of using technology to deliver business results, for example; using emerging innovations like Intelligent Transportation Systems and DOT's Connected Vehicles Program [7] designed to leverage vehicle-to-vehicle (V2V), driverless cars, and vehicle-to-infrastructure (V2I) wireless communication in order to make driving safer by making cars, trucks, buses and other vehicles aware of the vehicles around them, even if the drivers aren't.

Career

Pre US DOT CIO

While studying at the American University, Pradhan researched and experimented in using Internet for business. This was the era when Internet browsers were yet to be invented and the Internet itself was primarily a research and education tool. This effort helped him launch one of the first online businesses in the early 1990s. While his online company did not succeed, this early experimentation was vital for him joining the Center for Innovative Technology (CIT) [3] and later becoming its Managing Director.

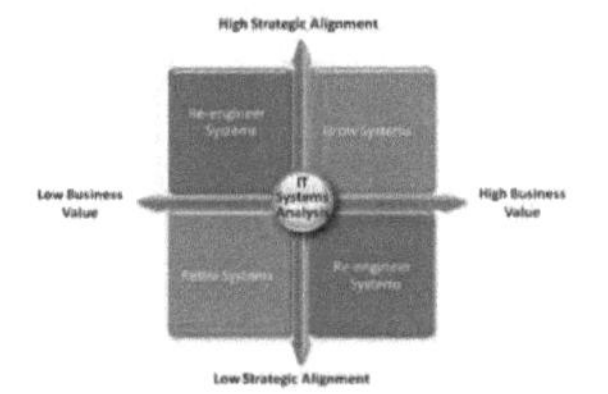

DOT IT Portfolio Rationalization

At CIT, Pradhan co-led in the creation and implementation of Virginia's first industry led Information Technology and Telecommunication's Strategic Plan in partnership with leading industry, university and government executives. He helped secure funding, established and expanded technology centers of excellence focused on increasing Internet utilization in partnership with Virginia universities, government and companies. He was also a mentor for entrepreneurs and startups seeking to build profitable companies around Internet technologies.

Pradhan also championed bold innovative initiatives launched in the DC metro region in the mid-nineties; including Potomac Knowledge Way [8] and Netpreneur,[9] established by the Morino Institute,[10] and the creation and expansion of the Northern Virginia Technology Council (NVTC) [11] a membership and trade association for the technology community in Northern Virginia, now the largest technology council in the nation, currently serving about 1,000 organizations. After CIT, Pradhan co-founded a wireless startup and was the CEO in early 2000.

Pradhan's interest in technology and its potential for positive impact on the education sector attracted him to join Fairfax County Public Schools (FCPS), the 12th largest school district in the USA. During his tenure, he promoted digital innovation and technology based solutions as an answer to educational challenges and opportunities, and helped transform the organization in to a premier technology integrated educational institution. He started innovative initiatives like the Technology@Home [12] which helped Fairfax County students, faculty and staff members purchase technology products directly from vendors at educational discount rates and Vendor Technology Briefings to match cutting edge educational technologies for FCPS needs. During this time, FCPS IT was listed tenth in top 100 best places to work [13] by Computer World magazine, and was the only K12 organization in the CIO 100 [14] list published by CIO magazine.

US Department of Transportation Chief Information Officer

In 2009, Pradhan joined DOT as its CIO. His focus at DOT has been on using technology to drive business value,[6] IT portfolio optimization,[15] streamlining technology services,[16] creating an IT business catalogue and Online distribution(app store),[5] holistic cyber security [17] [18] and public private partnerships like the Digital Transportation Ecosystem (DTE).[19] Some of the DOT technology initiatives include:

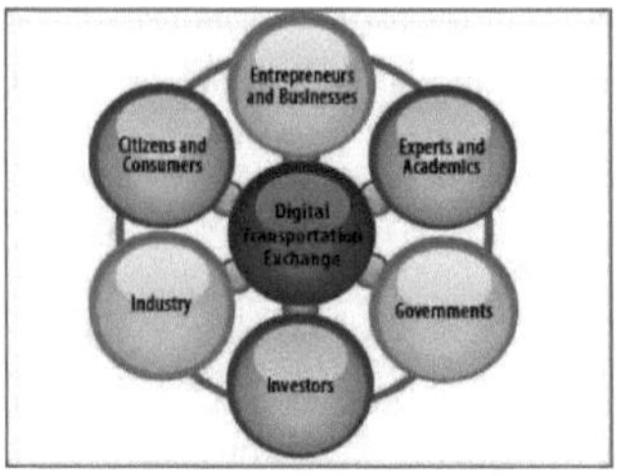

Digital Transportation Exchange (DTE)

- Modernization of the National Airspace System (NAS)-[20] Called NextGEN, which upgrades the system from older ground based radars to satellite and GPS based systems.
- Intelligent Transportation System (ITS)-[21] A Connected Vehicles Program that improves vehicle-to-vehicle and vehicle-to-infrastructure safety via wireless communications and threat alerts.
- Modernizing Analog 911 Systems (E911)-[22] Next generation digital VOIP 911 systems capable of receiving and sending internet protocol based voice, video and text.

Professional recognition

At the U.S. Department of Transportation, Pradhan earned the prestigious national InformationWeek's The Government CIO 50: Driving Change In The Public Sector national award [23] for bringing a business person's point of view to management of the DOT's IT strategy, policy, and implementation. Under Pradhan's leadership, US Department of Transportation IT has also earned InformationWeek 500 IT Innovators Awards,[24] including one for an Open Government initiative and the other for FAA's NextGen. They also received the White House Open Government Leading Practices Awards [25] designation in two of the four award categories, including "Leadership, Governance, Culture Change" and "Flagship Initiative."

ComputerWorld magazine named Pradhan to Premier 100 award [26] list where he joined an elite group of top US IT executives. The Premier 100 was created in 2000 to spotlight individuals who have made a positive impact on their organization through their work in technology. Each year, 100 honorees are selected to receive this lifetime recognition award.[27]

External links

- United States Department of Transportation (DOT) [28]
- DOT Office of the CIO [29]
- Nitin Pradhan US DOT I CIO.GOV Bio [30]
- US DOT IT Spending - DOT [31]
- Information Week: The Government CIO 50 - Driving Change In The Public Sector [32]
- The White House Leading Practices Awards for Open Government - US DOT Winner [33]
- ComputerWorld Premier 100 Honoree [34]
- Winning Through Technology Blog [35]

References

[1] http://www.dot.gov/cio/index.html/

[2] "Federal IT Dashboardl US DOT IT Portfolio" (http://www.itdashboard.gov/?q=portfolios/agency=021/)

[3] "Center for Innovative Technology" (http://www.cit.org/)

[4] "Nitin Pradhan's Personal Blog "Winning Through Technology"" (http://winningthroughtechnology.com/)

[5] "Making the Web Work for DOT: IdeaHub" (https://www.cio.gov/pages.cfm/page/Making-the-Web-Work-for-DOT-IdeaHub/)

[6] "The Changing Role of the CIO" (http://www.cio.gov/pages.cfm/page/The-Changing-Role-of-the-CIO/)

[7] "IT Solutions – Connected Vehicle Technology" (http://winningthroughtechnology.com/2011/09/03/connected-vehicle-technology/)

[8] "Potomac Knowledge Way (archived site)" (http://knowledgeway.org/)

[9] "Netpreneur (archived site)" (http://netpreneur.org/)

[10] "Morino Institute" (http://www.morino.org/)

[11] "Northern Virginia Technology Council (NVTC)" (http://www.nvtc.org/index.php/)

[12] "Technology@Home" (http://www.fcps.edu/it/technology/TechnologyHome.shtml/)

[13] "100 Best Places to Work in IT Companies in Information Technology" (http://www.computerworld.com/s/article/9024401/100_Best_Places_to_Work_in_IT_Company_Profile_Information?pageNumber=10/)

[14] "CIO 100 for 2007" (http://www.cio.com/cio100/detail/1695/)

[15] "IT Portfolio Rationalization – Effective Optimization of IT Funds" (http://winningthroughtechnology.com/2011/11/05/portfolio-rationalization-effective-optimization-of-it-funds/)

[16] "Telephony – Streamlining Service delivery" (http://winningthroughtechnology.com/2011/10/26/telephony-streamlining-service-delivery/)

[17] "The Business of Cyber security" (http://winningthroughtechnology.com/2011/10/08/the-âbusinessâ-of-cybersecurity/)

[18] "DOT CIO Questions FISMA Audits' Value" (http://www.govinfosecurity.com/articles.php?art_id=3125&rf=2010-11-24-eg/)

[19] "Transportation Dept. Creates Online Technology Exchange" (http://www.informationweek.com/news/government/leadership/231500483/)

[20] "NextGEN National Aerospace System" (http://www.faa.gov/nextgen/)

[21] "Intelligent Transportation Systems" (http://www.its.dot.gov/)

[22] "Enhanced 911 System" (http://www.its.dot.gov/ng911/)

[23] "InformationWeek's The Government CIO 50" (http://www.informationweek.com/galleries/showImage.jhtml?galleryID=348&imageID=34&articleID=220200135//)

[24] "InformationWeek's 500 Award Government IT Innovators" (http://www.informationweek.com/1278/)

[25] "White House Open Government Award" (http://www.whitehouse.gov/blog/2010/08/12/race-top-openness-and-innovation-announcing-agency-open-government-plan-leading-prac/)

[26] "ComputerWorld Premier 100 Lifetime Achievement Award" (http://www.computerworld.com/spring/p100/detail/379/)

[27] "Computerworld Research" (http://www.computerworld.com/s/article/9064798/Computerworld_Research#p100/)

[28] http://www.dot.gov/

[29] http://www.dot.gov/cio/

[30] http://www.cio.gov/profile/Pradhan-Nitin.cfm/

[31] http://it.usaspending.gov/?q=portfolios/agency=021/

[32] http://www.informationweek.com/galleries/showImage.jhtml?galleryID=348&imageID=34&articleID=220200135/

[33] http://www.whitehouse.gov/blog/2010/08/12/race-top-openness-and-innovation-announcing-agency-open-government-plan-leading-prac/

[34] http://www.computerworld.com/spring/p100/detail/379

[35] http://winningthroughtechnology.com/

Chief_information_officer

Chief information officer (CIO), or **information technology (IT) director**, is a job title commonly given to the most senior executive in an enterprise responsible for the information technology and computer systems that support enterprise goals. The title of Chief Information Officer in Higher Education may be the highest ranking technology executive although depending on the institution, alternative titles are used to represent this position. Generally, the CIO reports to the chief executive officer, chief operations officer or chief financial officer. In military organizations, they report to the commanding officer.

Quote From a CIO

I am always asked for my advice on how I am leading a large and sizeable Information Technology department in these turbulent economic times, in practical terms, a CIO (IT) has to be as much a business process manager as a technology manager. IT affects virtually every aspect of company operations today. Because of this, understanding where technology can improve a business process, streamline operations and consolidate functionality and resources across business units, or get new customers, for example, is vital. In order to understand this, today's TOP IT Executives like me have to know where the company gets its parts and pieces, and how those things are processed into a product or service someone will buy.

I must have command of what are the basic business processes and how they map to IT. I take a leadership position around helping the organization optimize its business processes; leveraging technology and being both champion, visionary and a leader but that requires good alignment between IT and rest of the business.

As a Top IT Exec you have to be "hungry for change" first. An IT Exec is uniquely positioned to be the major business transformation driver for his/her enterprises. We have a special visibility across the enterprise with an understanding of the core processes and information that make the organization tick. We need to be a catalyst for change across the enterprise and, change will require flexible, extendible IT infrastructures to respond to rapid and constant change. So, approaches like SOA, open standards, virtualization and consolidation, converged communications, cloud computing, multi-sourcing and shared services gain new appeal in this environment. Change also brings risk and our jobs are all about managing and minimizing risks so, enhanced risk mitigation, business resiliency, security, and service management initiatives will be a key to success.

Ramin Rastin Chief Information Officer & Regional Information Technology Vice President

CIO

Information technology and its systems have become so important that the CIO has come to be viewed in many organizations as the key contributor in formulating strategic goals for an organization. The CIO manages the implementation of the useful technology to increase information accessibility and integrated systems management. As a comparison, where the CIO adapts systems through the use of existing technologies, chief technology officer develops new technologies to expand corporate technological capabilities. When both positions are present in an organization, the CIO is generally responsible for processes and practices supporting the flow of information, whereas the CTO is generally responsible for technology infrastructure.

CIO magazine's "State of the CIO 2008" survey asked 558 IT leaders who they report to. The results were: CEO (41%), CFO (23%), COO (16%), Corporate CIO (7%) and Other (13%).[1]

Information technology

The prominence of the CIO position has risen greatly as information, and the information technology that drives it, has become an increasingly important part of the modern organization. The CIO may be a member of the executive committee of an organization, and/or may often be required to engage at board level depending on the nature of the organization and its operating structure and governance environment. No specific qualification are intrinsic of the CIO position, though the typical candidate may have expertise in a number of technological fields - computer science, software engineering, or information systems. Many candidates have Master of Business Administration or Master of Science in Management degrees.[2] More recently CIOs' leadership capabilities, business acumen and strategic perspectives have taken precedence over technical skills. It is now quite common for CIOs to be appointed from the business side of the organization, especially if they have project management skills.

In 2007 a survey amongst CIOs by *CIO* magazine in the UK discovered that their top 10 concerns were: people leadership, managing budgets, business alignment, infrastructure refresh, security, compliance, resource management, managing customers, managing change and board politics.[3]

In 2010, Gartner Executive Programs conducted a global CIO survey and received responses from 2,014 CIOs from 50 countries and 38 industries. [4] Gartner reported that the survey results indicated that the top ten technology priorities for CIOs for 2011 were cloud computing, virtualization, mobile technologies, IT management, business intelligence, networking, voice and data communications, enterprise applications, collaboration technologies, infrastructure, and Web 2.0.

Typically, a CIO is involved with driving the analysis and re-engineering of existing business processes, identifying and developing the capability to use new tools, reshaping the enterprise's physical infrastructure and network access, and with identifying and exploiting the enterprise's knowledge resources. Many CIOs head the enterprise's efforts to integrate the Internet into both its long-term strategy and its immediate business plans. CIO's are often tasked with either driving or heading up crucial IT projects which are essential to the strategic and operational objectives of an organisation. A good example of this would be the implementation of an Enterprise Resource Planning (ERP) system which typically has wide-ranging implications for most organizations. The CIO is evolving into a role where he/she is creating and monitoring business value from IT assets, to the point where corporate strategist Chris Potts suggests in the novel FruITion that the Chief Information Officer (CIO) be replaced with Chief Internal Investments Officer (CIIO).[5]

Another way that the CIO role is changing is an increased focus on service management.[6] As SaaS, IaaS, BPO and other more flexible value delivery techniques are brought into organizations the CIO usually functions as a 3rd party manager for the organization. In essence, a CIO in the modern organization is required to possess business skills and the ability to relate to the organization as a whole, as opposed to being a technological expert with limited functional business expertise. The CIO position is as much about anticipating trends in the market place with regards to technology as it is about ensuring that the business navigates these trends through expert guidance and proper strategic IT planning that is aligned to the corporate strategy of the organization.

See also

- Fractional CIO
- Chief Information Officer in Higher Education
- Chief technology officer
- IT Governance
- IT Strategy
- Ramin Rastin, CIO

References

[1] "State of the CIO 2008 Data Shows CIO Salaries, Influence Rising" (http://www.cio.com/article/147950/_State_of_the_CIO_2008_Data_Shows_CIO_Salaries_Influence_Rising). CIO. . Retrieved 27 February 2010.

[2] Meridith Levinson (2007-07-05). "Should You Get an MBA? - CIO.com - Business Technology Leadership" (http://www.cio.com/article/122507/Should_You_Get_an_MBA_). CIO.com. . Retrieved 2012-03-28.

[3] "Granger: The final word - CIO UK Magazine" (http://www.cio.co.uk/concern/budgets/features/index.cfm?articleid=351). Cio.co.uk. 2012-03-14. . Retrieved 2012-03-28.

[4] "Gartner Executive Programs Worldwide Survey of More Than 2,000 CIOs Identifies Cloud Computing as Top Technology Priority for CIOs in 2011" (http://www.gartner.com/it/page.jsp?id=1526414). Gartner. . Retrieved 23 March 2011.

[5] fruITion: Creating the Ultimate Corporate Strategy for Information Technology, Chris Potts, Technics Publications, LLC 2008 (http://www.technicspub.com/product.sc?productId=7&categoryId=1)

[6] "CIO Magazine: Recession Shifts IT Service Management into Fast Lane" (http://www.cio.com/article/558564/Recession_Shifts_IT_Service_Management_Into_Fast_Lane). Cio.com. 2010-02-26. . Retrieved 2012-03-28.

- CIOWikipedia.com (http://www.CIOWikipedia.com)
- Chief-information-officer.info (http://www.chief-information-officer.info)
- Ramin Rastin Chief information officer (http://www.raminrastin.com)
- US Federal CIO Council (http://www.cio.gov)
- UK Government - CIO Council (http://www.cabinetoffice.gov.uk/cio.aspx) Dead link as on 4th Nov 2010, unable to find new official site.
- The Chief Information Officer Concept in E-government: Lessons for Developing Countries. (http://esaconf.un.org/WB/default.asp?action=9&boardid=10&read=3538&fid=97) By D.C. Misra. On United Nations Department of Economic and Social Affairs WebBoard.

United_States_Department_of_Transportation

United States Department of Transportation	
Seal of the Department of Transportation	
Department overview	
Formed	April 1, 1967
Jurisdiction	United States of America
Headquarters	1200 New Jersey Avenue SE, Washington, D.C.
Employees	58,622
Annual budget	$79 billion USD (FY2011), *requested*[1]
Department executives	Ray LaHood, Secretary of Transportation John Porcari, Deputy Secretary of Transportation Joan DeBoer, Chief of Staff Robert S. Rivkin, General Counsel Nitin Pradhan, CIO Chris Bertram, CFO Vacant, Under Secretary for Policy
Child agencies	Federal Aviation Administration Federal Highway Administration Federal Railroad Administration Federal Transit Administration Maritime Administration Additional agencies
Website	
www.dot.gov [2]	

The **United States Department of Transportation** (**USDOT** or **DOT**) is a federal Cabinet department of the United States government concerned with transportation. It was established by an act of Congress on October 15, 1966, and began operation on April 1, 1967. It is administered by the United States Secretary of Transportation.

Its mission is to "Serve the United States by ensuring a fast, safe, efficient, accessible, and convenient transportation system that meets our vital national interests and enhances the quality of life of the American people, today and into the future."

History

Prior to the Department of Transportation, the Under Secretary of Commerce for Transportation administered the functions now associated with the DOT. In 1965, Najeeb Halaby, administrator of the Federal Aviation Administration (FAA), suggested to President Lyndon B. Johnson that transportation be elevated to a cabinet-level post, and that the FAA be folded into the DOT.

Agencies

- Federal Aviation Administration (FAA)
- Federal Highway Administration (FHWA)
- Federal Motor Carrier Safety Administration (FMCSA)
- Federal Railroad Administration (FRA)
- Federal Transit Administration (FTA)
- Maritime Administration (MARAD)
- National Highway Traffic Safety Administration (NHTSA)
- Office of Inspector General (OIG)
- Office of the Secretary of Transportation (OST)
- Pipeline and Hazardous Materials Safety Administration (PHMSA)
- Research and Innovative Technology Administration (RITA)
- Saint Lawrence Seaway Development Corporation (SLSDC)
- Surface Transportation Board (STB)

New headquarters building for the U.S. Department of Transportation, which opened in Spring 2007 in Southeast Washington, near the Washington Navy Yard and Nationals Park

Former Agencies

- Transportation Security Administration – transferred to Department of Homeland Security in 2003
- United States Coast Guard – transferred to Department of Homeland Security in 2003

The seal of the US Department of Transportation prior to 1980

Related legislation

- 1806 – Cumberland Road
- 1862 – Pacific Railway Act
- 1887 – Interstate Commerce Act
- 1916 – Adamson Railway Labor Act
- 1935 – Motor Carrier Act
- 1946 – Federal Airport Act PL 79-377
- 1950 – Federal Aid to Highway PL 81-769

- 1954 – Saint Lawrence Seaway Act
- 1956 – Federal-Aid to Highway/Interstate Highway Act PL 84-627
- 1957 – Airways Modernization Act PL 85-133
- 1958 – Transportation Act PL 85-625
- 1958 – Federal Aviation Act PL 85-726
- 1959 – Airport Construction Act PL 86-72
- 1964 – Urban Mass Transportation Act PL 88-365
- 1965 – Highway Beautification Act PL 89-285
- 1966 – Department of Transportation established PL 89-670
  Uniform Time Act
- 1970 – Urban Mass Transportation Act PL 91-453
- 1970 – Rail Passenger Service Act PL 91-518
- 1970 – Airport and Airway Development Act PL 91-258
- 1973 – Federal Aid Highway Act PL 93-87
- 1973 – Amtrak Improvement Act PL 93-146
- 1973 – Federal Aid Highway Act PL 93-87
- 1974 – National Mass Transportation Assistance Act PL 93-503
- 1976 – Railroad Revitalization and Regulatory Reform Act PL 94-210
- 1976 – Hart-Scott-Rodino Antitrust Improvements Act PL 94-435
- 1978 – Airline Deregulation Act PL 95-504
- 1980 – Motor Carrier Act PL 96-296
- 1980 – Staggers Rail Act PL 96-448
- 1982 – Transportation Assistance Act PL 97-424
- 1982 – Bus Regulatory Reform Act PL 97-261
- 1987 – Surface Transportation Act PL 100-17
- 1991 – Intermodal Surface Transportation Efficiency Act PL 102-240
- 1998 – Transportation Equity Act for the 21st Century PL 105-178
- 2000 – Wendell H. Ford Aviation Investment and Reform Act for the 21st Century PL 106-181
- 2002 – Homeland Security Act (PL 107-296)
- 2005 – Safe, Accountable, Flexible, Efficient Transportation Equity Act: A Legacy for Users (PL 109-59)

Budget

The DOT will award $742.5 million in funds from the American Recovery and Reinvestment Act to 11 transit projects. The awardees include light rail projects. Other projects include both a commuter rail extension and a subway project in New York, New York, and a bus rapid transit system in Springfield, Oregon. The funds subsidize a heavy rail project in northern Virginia, completing the Washington Metropolitan Area Transit Authority's Metro Silver Line to connect Washington, D.C., and the Washington Dulles International Airport.[3] (DOT had previously agreed to subsidize the Silver Line construction to Reston, Virginia.)[4]

President Barack Obama's budget request for fiscal year 2010 also includes $1.83 billion in funding for major transit projects, of which more than $600 million will go towards 10 new or expanding transit projects. The proposed budget provides additional funding for all of the projects currently receiving Recovery Act funding, except for the bus rapid transit project. It also continues funding for another 18 transit projects that are either currently under construction or soon will be.[3]

See also

- National Transportation Safety Board
- Passenger vehicles in the United States
- Transportation in the United States
- United States Federal Maritime Commission
- United States
- Ad Council

References

[1] "2011 Budget Hightlights" (http://www.dot.gov/budget/2011/2011budgethighlights.pdf). . Retrieved 2010-09-30.
[2] http://www.dot.gov
[3] "DOT Awards $742.5 Million in Recovery Act Funds to 11 Transit Projects" (http://apps1.eere.energy.gov/news/news_detail.cfm/news_id=12512). *EERE Network News*. May 13, 2009. . Retrieved 2010-08-09.
[4] "Annual Report on Funding Recommendations – Fiscal Year 2010" (http://www.fta.dot.gov/documents/20090508_Release_FY_2010_Annual_Report.pdf) (PDF). U.S. Department of Transportation. 29 April 2009. pp. A-75 (101) & seq. . Retrieved 2010-08-09.

External links

- United States Department of Transportation Official Website. (http://www.dot.gov/)
- United States Department of Transportation – Office of the Historian. (http://dotlibrary.dot.gov/Historian/historian.htm)
- Proposed and final federal regulations from the Department Of Transportation (http://openregs.com/agencies/view/1/)
- Papers of Louis S. Rothschild (Under Secretary of Commerce for Transportation 1955–1958), Dwight D. Eisenhower Presidential Library (http://www.eisenhower.archives.gov/Research/Finding_Aids/R.html)
- United States Department of Transportation (http://www.whorunsgov.com/institutions/transportation/) at *WhoRunsGov* at *The Washington Post*
- Automotive/Highway Transportation Reports & Documents (http://www.carsandracingstuff.com/library/reports/reports.php) at The Crittenden Automotive Library (most from FHWA/NHTSA)

Presidency_of_Barack_Obama

Barack Obama		
[[file:Official portrait of Barack Obama.jpg		alt=]]
44th President of the United States		
Incumbent		
Assumed office January 20, 2009		
Vice President	Joe Biden	
Preceded by	George W. Bush	
Personal details		
Born	Barack Hussein Obama II[1] August 4, 1961Honolulu, Hawaii, United States[1]	
Nationality	American	
Political party	Democratic	
Spouse(s)	Michelle Robinson (m. 1992)	
Children	Malia (b.1998) Sasha (b.2001)	
Residence	The White House	
Alma mater	Occidental College Columbia University (B.A.) Harvard Law School (J.D.)	
Profession	Community organizer Attorney Author Constitutional law professor United States Senator President of the United States	
Religion	Christian,[2] former member of United Church of Christ[3] [4]	
Signature		
Website	WhiteHouse.gov [5]	

The **Presidency of Barack Obama** began at noon EST on January 20, 2009, when he became the 44th President of the United States. Obama was a United States Senator from Illinois at the time of his victory over Arizona Senator John McCain in the 2008 presidential election. Barack Obama is the first African-American president of the United States, as well as the first born in Hawaii.

Barack Obama being sworn in as President of the United States.

His policy decisions have addressed a global financial crisis and have included changes in tax policies, legislation to reform the United States health care industry, foreign policy initiatives and the phasing out of detention of prisoners at the Guantanamo Bay detention camp in Cuba.

He attended the G-20 London summit and later visited U.S. troops in Iraq. On the tour of various European countries following the G-20 summit, he announced in Prague that he intended to negotiate substantial reduction in the world's nuclear arsenals, en route to their eventual extinction. In October 2009, Obama was awarded the Nobel Peace Prize for "his extraordinary efforts to strengthen international diplomacy and cooperation between peoples."

Transition period

The presidential transition period began following Obama's election to the presidency on November 4, 2008. The Obama-Biden Transition Project was co-chaired by John Podesta, Valerie Jarrett, and Pete Rouse. During the transition period, Obama announced his nominations for his Cabinet and administration. Shortly after the election on November 4, Obama chose Representative Rahm Emanuel of Illinois as White House Chief of Staff.[6]

Cabinet nominations included former Democratic primary opponents Hillary Rodham Clinton for Secretary of State and Bill Richardson for Secretary of Commerce (although the latter withdrew on January 4, 2009). Obama appointed Eric Holder as his Attorney General, the first African-American appointed to that position. He also nominated Timothy F. Geithner to serve as Secretary of the Treasury.[7] On December 1, Obama announced that he had asked Robert Gates to remain as Secretary of Defense, making Gates the first Defense head to carry over from a president of a different party.[8] He nominated former Assistant Secretary of State for African Affairs Susan Rice to the United States Ambassador to the United Nations, which he restored to a Cabinet-level position.[9]

During his transition, he maintained a website Change.gov, on which he wrote blogs to readers and uploaded video addresses by many of the members of his new cabinet.[10] He announced strict rules for federal lobbyists, restricting them from financially contributing to his administration and forcing them to stop lobbying while working for him.[11] The website also allowed individuals to share stories and visions with each other and the transition team in what was called the Citizen's Briefing Book, which was given to Obama shortly after his inauguration.[12] Most of the information from Change.gov was transferred to the official White House website whitehouse.gov just after Obama's inauguration.[13]

Inauguration

Barack Obama was inaugurated on January 20, 2009. He officially assumed the presidency at 12:00 noon, EST,[14] and completed the oath of office at 12:05 PM, EST. He delivered his inaugural address immediately following his oath. After his speech, he went to the President's Room in the House Wing of the Capitol and signed three documents: a commemorative proclamation, a list of Cabinet appointments, and a list of sub-Cabinet appointments, before attending a luncheon with congressional and administration leaders and invited guests.[15] To commemorate the 200th anniversary of the birth of former President Abraham Lincoln, the same Bible that was used for Lincoln's inauguration was used in Obama's inauguration.[16]

In administering the oath, Chief Justice John G. Roberts misplaced the word "faithfully" and erroneously replaced the phrase "President of the United States" with "President to the United States" before restating the phrase correctly; since Obama initially repeated the incorrect form, some scholars argued the President should take the oath again.[17] On January 21, Roberts readministered the oath to Obama in a private ceremony in the White House Map Room, making him the seventh U.S. president to retake the oath; White House Counsel Greg Craig said Obama took the oath from Roberts a second time out of an "abundance of caution".[18]

First 100 days

Expectations

Obama's 100th day in office was April 29, 2009. In his first post-election interview with *60 Minutes*, Obama said that he has been studying Franklin Roosevelt's first 100 days,[19] while adding, "The first hundred days is going to be important, but it's probably going to be the first thousand days that makes the difference."[20]

Obama's first 100 days were highly anticipated ever since he became the presumptive nominee.[21] Several news outlets created web pages dedicated to covering the subject.[22] Commentators weighed in on challenges and priorities within domestic, foreign, economic, and environmental policy.[23] [24] [25] [26] CNN lists a number of economic issues that "Obama and his team will have to tackle in their first 100 days", foremost among which is passing and implementing a recovery package to deal with the financial crisis.[25] Clive Stafford Smith, a British human rights lawyer, expressed hopes that the new president will close Guantanamo Bay detention camp in his first 100 days in office.[24] After aides of the president announced his intention to give a major foreign policy speech in the capital of an Islamic country, there were speculations in Jakarta that he might return to his former home city within the first 100 days.[27]

The New York Times devoted a five-part series, which was spread out over two weeks, to anticipatory analysis of Obama's first hundred days. Each day, the analysis of a political expert was followed by freely edited blog postings from readers. The writers compared Obama's prospects with the situations of Franklin D. Roosevelt (January 16, Jean Edward Smith),[28] John F. Kennedy (January 19, Richard Reeves),[29] Lyndon B. Johnson (January 23, Robert Dallek),[30] Ronald Reagan (January 27, Lou Cannon),[31] and Richard Nixon.

Legislation and executive orders

Within minutes of taking the Oath of Office on January 20, Obama's Chief of Staff, Rahm Emanuel, issued an order suspending last-minute federal regulations pushed through by outgoing President George W. Bush, planning to review everything still pending.[32] Due to the economic crisis, the President enacted a pay freeze for Senior White House Staff making more than $100,000 per year,[33] as well as announcing stricter guidelines regarding lobbyists in an effort to raise the ethical standards of the White House.[34] He asked for a waiver to his own new rules, however, for the appointments of William Lynn to the position of Deputy Defense Secretary, Jocelyn Frye to the position of director of policy and projects in the Office of the First Lady, and Cecilia Muñoz to the position of director of intergovernmental affairs in the executive office of the president, leading to some criticism of hypocrisy and violation of his pledge for governmental openness.[35] [36]

In his first week in office, Obama signed Executive Order 13492 suspending all the ongoing proceedings of Guantanamo military commission and ordering the detention facility to be shut down within the year.[37] [38] [39] He also signed Executive Order 13491 - Ensuring Lawful Interrogations requiring the Army Field Manual to be used as a guide for terror interrogations, banning torture and other coercive techniques, such as waterboarding.[40] Obama also issued an executive order entitled "Ethics Commitments by Executive Branch Personnel", setting stricter limitations on incoming executive branch employees and placing tighter restrictions on lobbying in the White House.[41] Obama signed two Presidential Memoranda concerning energy independence, ordering the Department of Transportation to establish higher fuel efficiency standards before 2011 models are released and allowing states to raise their emissions standards above the national standard.[42] He also ended the Mexico City Policy, which banned federal grants to international groups that provide abortion services or counseling.[43] [44]

In his first week he also established a policy of producing a weekly Saturday morning video address available on whitehouse.gov and YouTube,[45] [46] [47] much like those released during his transition period.[48] [49] The first address had been viewed by 600,000 YouTube viewers by the next afternoon.[50]

The first piece of legislation Obama signed was the Lilly Ledbetter Fair Pay Act of 2009 on January 29, which revised the statute of limitations for filing pay discrimination lawsuits. Lilly Ledbetter joined Obama and his wife,

Michelle, as he signed the bill, fulfilling his campaign pledge to nullify *Ledbetter v. Goodyear*.[51] On February 3, he signed the Children's Health Insurance Program Reauthorization Act (CHIP), expanding health care from 7 million children under the plan to 11 million.[52]

After much debate, the American Recovery and Reinvestment Act (ARRA) was passed by both the House and Senate on February 13, 2009. Originally intended to be a bipartisan bill, the passage of the bill was largely along party lines. No Republicans voted for it in the House, and three moderate Republicans voted for it in the Senate (Susan Collins and Olympia Snowe of Maine and Arlen Specter of Pennsylvania).[53] The bill combined tax breaks with spending on infrastructure projects, extension of welfare benefits, and education.[54] [55] The final cost of the bill was $787 billion, and almost $1.2 trillion with debt service included.[56] Obama signed the Act into law on February 17, 2009, in Denver, Colorado.[57]

On March 9, 2009, Obama lifted restrictions on federal funding of embryonic stem cell research,[58] and in doing so, called into question some of George W. Bush's signing statements. Obama stated that he too would employ signing statements if he deems upon review that a portion of a bill is unconstitutional,[59] [60] and he has issued several signing statements.[61]

Early in his presidency, Obama signed a law raising the tobacco tax 62 cents on a pack of cigarettes.[62] The tax is to be "used to finance a major expansion of health insurance for children", and "help some [smokers] to quit and persuade young people not to start".[62]

In October 2011, Obama instituted the We Can't Wait program, which involved using executive orders, administrative rulemaking, and recess appointments to institute policies without the support of Congress.[63] The initiative was developed in response to Congress's unwillingness to pass economic legislation proposed by Obama, and conflicts in Congress during the 2011 debt ceiling crisis.[64]

Approval ratings and opinion

2009

After his transition period, Obama entered office with an approval rating of 82%.[65] At the end of his first week, 68% of respondents in a Gallup poll approved of how Obama was handling his job, the second highest approval rating for a President shortly after being elected since World War II.[66] Throughout early February polls showed scattered approval ratings: 62% (CBS News),[67] 64% (USA Today/Gallup), 66% (Gallup), and 76% in an outlier poll (CNN/Opinion Research).[68] [69] Gallup reported the congressional address in late February boosted his approval from a term-low of 59% to 67%.[70]

Throughout autumn 2009, Rasmussen estimated Obama's approval as fluctuating between 45% and 52% and his disapproval between 48% and 54%;[71] as of November 11, Pew Research estimated Obama's approval between 51% and 55% and his disapproval between 33% and 37% since July.[72]

Handling of the economy

Rasmussen reported in mid-February 2009, that 55% of voters gave Obama good or excellent marks on his handling of the economy.[73] In early March, a *The Wall Street Journal* survey of 49 economists gave Obama an average grade of 59 out of 100, with 42% of the respondents surveyed giving the administration's economic policies a grade below 60 percent. In comparison, only 30% of those same economists considered the response of governments around the world to the global recession to have been adequate.[74] In April, a Gallup poll showed trust in Obama's economic policy with 71% saying they had "a fair amount" or "a great deal" of confidence in Obama's handling of the economy, higher than for Ben Bernanke, Tim Geithner, or leaders of Congress.[75] Another Gallup poll in June showed 55% of Americans approved Obama's overall handling of the economy, but 48% and 51% disapproved of his handling of the federal budget deficit and controlling federal spending, respectively.[76] A CBS News poll taken August 27–31 showed 53% of those polled approved of his handling of the economy.[77] A Rasmussen poll taken on

November 12 found 45% of Americans rating Obama's handling of the economy as poor and 39% rating him as doing a good or excellent job. They found 72% of Democrats rated his handling of the economy as good or excellent, while only 10% of Republicans and 27% of voters not affiliated with either party agreed.[78]

2010

On March 25, 2010, following his signing of landmark health care reform legislation into law, Obama's polling was revealed by Bloomberg to be 50%, with higher marks for relations with other countries (58%) and his running of the war in Afghanistan (54%). "Obama's approval rating is roughly equal to what Bill Clinton had at this point in his presidency, according to data maintained by Gallup (and) higher than the 45 percent Ronald Reagan recorded in April 1982" and more favorable than Democrats or Republicans office in 2010. They found Obama's approval rating was at 85% among Democrats, compared with 46% among independents and 11% among Republicans.[79]

Fox News released the results of two polls on April 8–9, 2010. The first showed a drop in Obama's approval rating to 43%, with 48% disapproving. In that poll, Democrats approved of Obama's performance 80–12%, while independents disapproved 49–38%.[80] The other poll, which concentrated on the economy, showed disapproval of Obama's handling of the economy by a 53–42% margin, with 62% saying they were dissatisfied with the handling of the federal deficit.[81] According to a Gallup Poll released April 10, 2010, President Obama had a 45% approval rating, with 48% disapproving.[82] In a poll from Rasmussen Reports, released April 10, 2010, 47% approved of the President's performance, while 53% disapproved.[83] [84]

2011

Obama's approval rating jumped to a high following the death of Osama Bin Laden on May 2, 2011. A GfK poll conducted May 5, 2011 found his approval rating to be 60%. During the debt ceiling debate in August 2011, Obama's approval rating dropped to the low-40s.[85]

Major legislation

2009

- January 29: Lilly Ledbetter Fair Pay Act
- February 4: Children's Health Insurance Reauthorization Act
- February 11: DTV Delay Act
- February 17: American Recovery and Reinvestment Act
- March 30: Omnibus Public Lands Management Act of 2009
- April 21: Edward M. Kennedy Serve America Act
- May 20: Fraud Enforcement and Recovery Act
- May 20: Helping Families Save Their Homes Act of 2009
- May 22: Credit Card Accountability, Responsibility, and Disclosure Act of 2009
- June 22: Family Smoking Prevention and Tobacco Control Act
- August 6: Cash For Clunkers Extension Act
- October 22: Veterans Health Care Budget Reform and Transparency Act
- October 28: Matthew Shepard and James Byrd Jr. Hate Crimes Prevention Act
- October 30: Ryan White HIV/AIDS Treatment Extension Act
- November 6: Worker, Homeownership, and Business Assistance Act of 2009

2010

- January 27: Emergency Aid to American Survivors of Haiti Earthquake Act
- March 4: Travel Promotion Act
- March 18: Hiring Incentives to Restore Employment Act (HIRE Act)
- March 23: Patient Protection and Affordable Care Act
- March 30: Health Care and Education Reconciliation Act of 2010
- May 5: Caregivers and Veterans Omnibus Health Services Act of 2010
- May 17: Daniel Pearl Freedom of the Press Act
- July 1: Comprehensive Iran Sanctions, Accountability, and Divestment Act
- July 21: Dodd-Frank Wall Street Reform and Consumer Protection Act
- July 22: Unemployment Compensation Extension Act of 2010
- July 22: Improper Payments Elimination and Recovery Act of 2010
- August 10: SPEECH Act
- September 27: Small Business Jobs and Credit Act of 2010
- December 9: Animal Crush Video Prohibition Act of 2010
- December 13: Healthy, Hunger-Free Kids Act of 2010
- December 17: Tax Relief, Unemployment Insurance Reauthorization, and Job Creation Act of 2010
- December 22: Don't Ask, Don't Tell Repeal Act of 2010

2011

- January 2: James Zadroga 9/11 Health and Compensation Act of 2010 (H.R. 847 [86])
- January 4: Food Safety and Modernization Act, Pub.L. 111-353 [87], H.R. 2751 [88]
- August 2: Budget Control Act of 2011
- October 21: United States-Colombia Free Trade Agreement Implemention Act
- October 21: United States-Panama Free Trade Agreement Implemention Act
- October 21: United States-South Korea Free Trade Agreement Implemention Act

Personnel

Cabinet appointees

The Obama Cabinet		
Office	**Name**	**Term**
President	**Barack Obama**	2009–present

Vice President	**Joe Biden**	2009–present
Secretary of State	**Hillary Clinton**	2009–present
Secretary of Treasury	**Timothy Geithner**	2009–present
Secretary of Defense	**Robert Gates***	2006–2011
	Leon Panetta	2011–present
Attorney General	**Eric Holder**	2009–present
Secretary of the Interior	**Ken Salazar**	2009–present
Secretary of Agriculture	**Tom Vilsack**	2009–present
Secretary of Commerce	**Gary Locke**	2009–2011
	John Bryson	2011–present
Secretary of Labor	**Hilda Solis**	2009–present
Secretary of Health and Human Services	**Kathleen Sebelius**	2009–present
Secretary of Education	**Arne Duncan**	2009–present
Secretary of Housing and Urban Development	**Shaun Donovan**	2009–present
Secretary of Transportation	**Ray LaHood**	2009–present
Secretary of Energy	**Steven Chu**	2009–present
Secretary of Veterans Affairs	**Eric Shinseki**	2009–present
Secretary of Homeland Security	**Janet Napolitano**	2009–present
Chief of Staff	**Rahm Emanuel**	2009–2010
	William Daley	2011–2012
	Jacob Lew	2012–present
Administrator of the Environmental Protection Agency	**Lisa Jackson**	2009–present
Director of the Office of Management and Budget	**Peter Orszag**	2009–2010
	Jacob Lew	2010–2012
	Jeffrey Zients**	2012–present
Ambassador to the United Nations	**Susan Rice**	2009–present
United States Trade Representative	**Ron Kirk**	2009–present
Retained from previous administration** *****Acting		

Twenty-two members of the Obama administration are either in the United States Cabinet (15) or are in positions considered to be Cabinet-level (7). The members of the Cabinet are the heads of the fifteen major departments (State, Defense, Justice, etc.), and the seven cabinet-level positions are the Vice President, White House Chief of Staff, Administrator of the Environmental Protection Agency, Director of the Office of Management and Budget, U.S. Trade Representative, Ambassador to the United Nations, and the Chair of the Council of Economic Advisers.[89] [90] Since Robert Gates was a member of the previous administration, his letter of resignation (a formality at the end of a President's term) was simply not accepted, and he did not need confirmation.[91] On January 19, 2009, Senate Democratic leaders requested fifteen of the twenty positions to be ratified by unanimous consent,[92] and seven gained unanimous confirmation by voice vote the next day: Ken Salazar, Steven Chu, Arne Duncan, Peter Orszag, Eric Shinseki, Tom Vilsack, and Janet Napolitano.[91] [93] On January 21, Obama presided over the swearing in of the seven unanimous nominees.[94] Later that day, the Senate confirmed Hillary Clinton by a 94–2 vote. On January 22, several more confirmations were approved unanimously: Susan E. Rice, Ray LaHood, Lisa P. Jackson, and Shaun Donovan.[95] On January 26, the Senate confirmed Geithner by a 60–34 margin.[96] [97]

At the conclusion of Obama's first week as President, Hilda Solis, Tom Daschle, Ron Kirk, and Eric Holder had yet to be confirmed, and there had been no second appointment for Secretary of Commerce.[97] Holder was confirmed by a vote of 75–21 on February 2,[98] and on February 3, Obama announced Senator Judd Gregg as his second nomination for Secretary of Commerce.[99] Daschle withdrew later that day amid controversy over his failure to pay income taxes and potential conflicts of interest related to the speaking fees he accepted from health care interests.[100] Solis was later confirmed by a vote of 80-17 on February 24,[101] and Ron Kirk was confirmed on March 18 by a 92-5 vote in the Senate.[102]

Gregg, who was the leading Republican negotiator and author of the TARP program in the Senate, after publication that he had a multi-million dollar investment in the Bank of America, on February 12, withdrew his nomination as Secretary of Commerce, citing "irresolvable conflicts" with President Obama and his staff over how to conduct the 2010 census and the American Recovery and Reinvestment Act of 2009.[103] [104] Former Washington governor Gary Locke was nominated on February 26[105] as Obama's third choice for Commerce Secretary and confirmed on March 24 by voice vote.[106]

On March 2, Obama introduced Kansas governor Kathleen Sebelius as his second choice for Secretary of Health and Human Services. He also introduced Nancy-Ann DeParle as head of the new White House Office of Health Reform, which he suggested would work closely with the Department of Health and Human Services.[107] [108] At the end of March, Sebelius was the only remaining Cabinet member yet to be confirmed.[106]

Six high-ranking cabinet nominees in the Obama administration had their confirmations delayed or rejected among reports that they did not pay all of their taxes, including Tom Daschle, Obama's original nominee for Health and Human Services Secretary, and Treasury Secretary Timothy Geithner.[109] Though Geithner was confirmed, and Senator Max Baucus, chairman of the Senate Finance Committee, thought Daschle would have been confirmed, Daschle withdrew his nomination on February 3.[100] Obama had nominated Nancy Killefer for the position of Chief Performance Officer, but Killefer also withdrew on February 3, citing unspecified problems with District of Columbia unemployment tax.[110] A senior administration official said that Killefer's tax issues dealt with household help.[110] Hilda Solis, Obama's nominee for Secretary of Labor, faced delayed confirmation hearings due to tax lien concerns pertaining to her husband's auto repair business,[111] but she was later confirmed on February 24.[101] While pundits puzzled over U.S. Trade Representative-designate Ron Kirk's failure to be confirmed by March 2009, it was reported on March 2 that Kirk owed over $10,000 in back taxes. Kirk agreed to pay them in exchange for Senate Finance Committee Chairman Max Baucus's aid in speeding up the confirmation process;[112] he was later confirmed on March 18.[102] On March 31, Kathleen Sebelius, Obama's nominee for Health and Human Services secretary, revealed in a letter to the Senate Finance Committee that her Certified Public Accountant found errors in her tax returns for years 2005-2007. She, along with her husband, paid more than $7,000 in back taxes, along with $878 in interest.[113]

Notable non-Cabinet positions

Appointees serve at the pleasure of the President and were nominated by Barack Obama except as noted.

- Chair of the Federal Deposit Insurance Corporation
  - **Sheila Bair**[1] (2009–2011)
- Special envoy for Afghanistan and Pakistan
  - **Richard Holbrooke** (2009–2010)
  - **Marc Grossman** (2011–present)
- Assistant to the President for National Security Affairs
  - Retired General **James L. Jones** (2009–2011)
  - **Thomas E. Donilon** (2011–present)
- Special envoy to the Middle East
  - **George J. Mitchell** (2009–2011)
  - **David Hale** (2011–present)
- Director of the Federal Bureau of Investigation
  - **Robert Mueller**[2] (2009–present)
- Director of the Central Intelligence Agency
  - **Leon Panetta** (2009–2011)
  - **Michael Morell** (Acting-2011)
  - **David Petraeus** (2011–present)
- Chair of the Council of Economic Advisers
  - **Christina Romer** (2009–2010)
  - **Austan Goolsbee** (2010–2011)
  - **Alan Krueger** (2011–present)
- Chair of the Securities and Exchange Commission
  - **Mary Schapiro** (2009–present)
- Assistant to the President for Economic Policy and Director of National Economic Council
  - **Lawrence Summers** (2009–2010)
  - **Gene Sperling** (2011–present)
- Chairman of the Economic Recovery Advisory Board
  - **Paul Volcker** (2009–2011)
  - *Replaced by Council on Jobs and Competitiveness in 2011*
- Chairperson of the Council on Jobs and Competitiveness
  - **Jeffrey R. Immelt** (2011–present)

[1] Appointed by George W. Bush in 2006 to a five-year term

[2] Appointed by George W. Bush in 2001 to a ten-year term

Judicial nominees

United States Supreme Court

Obama appointed the following Justices to the Supreme Court of the United States:

- **Sonia Sotomayor** – 2009
- **Elena Kagan** – 2010

Further information: List of nominations to the Supreme Court of the United States

Other courts

Outside of the Supreme Court, by October 2009, Obama had nominated fewer than two dozen judges to fill judicial vacancies, of which there were close to 100. This has prompted some Democrats to criticize the pace of Obama's judicial appointments as too slow.[114] In December 2009, Senator Patrick Leahy criticized Republicans for stalling those judicial nominations that had been made, noting that the Senate confirmed more district and circuit court nominees during the first year of the George W. Bush administration than it had approved by that point during the Barack Obama presidency.[115]

As of July 2010, Obama's nominees to the district and circuit courts had been confirmed at a rate of only 43.5 percent, compared to 87.2 percent during Bill Clinton's administration and 91.3 percent for George W. Bush. The Center for American Progress, which compiled the data, commented:

> Judicial confirmations slowed to a trickle on the day President Barack Obama took office. Filibusters, anonymous holds, and other obstructionary tactics have become the rule. Uncontroversial nominees wait months for a floor vote, and even district court nominees—low-ranking judges whose confirmations have never been controversial in the past—are routinely filibustered into oblivion. Nominations grind to a halt in many cases even after the Senate Judiciary Committee has unanimously endorsed a nominee. [116]

Policies

Economy

Upon entering office, Obama planned to center his attention on handling the global financial crisis.[117] Even before his inauguration he lobbied Congress to pass an economic stimulus bill,[118] which became the top priority during his first month in office.[119] As President, Obama made a high profile trip to Capitol Hill, Washington, D.C. to dialog with Congressional Republicans and advocate for the bill.[120] On February 17, 2009, Obama signed into law a $787 billion plan that included spending for health care, infrastructure, education, various tax breaks and incentives, and direct assistance to individuals.[121] The tax provisions of the law reduced taxes for 98 percent of taxpayers, bringing tax rates to their lowest levels in 60 years.[122]

As part of the 2010 budget proposal, the Obama administration has proposed additional measures to attempt to stabilize the economy, including a $2–3 trillion measure aimed at stabilizing the financial system and freeing up credit. The program includes up to $1 trillion to buy toxic bank assets, an additional $1 trillion to expand a federal consumer loan program, and the $350 billion left in the Troubled Assets Relief Program. The plan also includes $50 billion intended to slow the wave of mortgage foreclosures.[121] The 2011 budget includes a three-year freeze on discretionary spending, proposes several program cancellations, and raises taxes on high income earners to bring down deficits during the economic recovery.[123]

Obama speaking with former President Bill Clinton and Senior Advisor Valerie Jarrett about job creation in July 2010.

In a July 2009 interview with ABC News, Biden was asked about the sustained increase of the U.S. unemployment rate from May 2007 to October 2009[124] despite the administration's multi-year economic stimulus package passed five months earlier. He responded "The truth is, we and everyone else, misread the economy. The figures we worked off of in January were the consensus figures and most of the blue chip indexes out there ... the truth is, there was a misreading of just how bad an economy we inherited."[125] [126] The White House indicates that 2 million jobs were created or saved due to the stimulus package in 2009[127] and self reporting by recipients of the grants, loans, and contracts portion of the package report that the package saved or created 608,317 jobs in the final three months of 2009.[128]

The unemployment rate rose in 2009, reaching a peak in October at 10.1% and averaging 10.0% in the fourth quarter.[129] Following a decrease to 9.7% in the first quarter of 2010, the unemployment rate fell to 9.6% in the second quarter, where it remained for the rest of the year.[129] Between February and December 2010, employment rose by 0.8%, which was less than the average of 1.9% experienced during comparable periods in the past four employment recoveries.[130] GDP growth returned in the third quarter of 2009, expanding at a 1.6% pace, followed by a 5.0% increase in the fourth quarter.[131] Growth continued in 2010, posting an increase of 3.7% in the first quarter, with lesser gains throughout the rest of the year.[131] Overall, the economy expanded at a rate of 2.9% in 2010.[132]

During November–December 2010, Obama and a lame duck session of the 111th Congress focused on a dispute about the temporary Bush tax cuts, which were due to expire at the end of the year. Obama wanted to extend the tax cuts for taxpayers making less than $250,000 a year. Congressional Republicans agreed but also wanted to extend the tax cuts for those making over that amount, and refused to support any bill that did not do so.[133] All the Republicans in the Senate also joined in saying that, until the tax dispute was resolved, they would filibuster to prevent consideration of any other legislation, except for bills to fund the U.S. government.[134] [135] On 7 December, Obama strongly defended a compromise agreement he had reached with the Republican congressional leadership that included a two-year extension of all the tax cuts, a 13-month extension of unemployment insurance, a one-year reduction in the FICA payroll tax, and other measures.[136] On December 10, Senator Bernie Sanders (I-VT) led a filibuster against the compromise tax proposal, which lasted over eight hours.[137] Obama persuaded many wary Democrats to support the bill,[138] but not all; of the 148 votes against the bill in the House, 112 were cast by Democrats and only 36 by Republicans.[139] The $858 billion Tax Relief, Unemployment Insurance Reauthorization, and Job Creation Act of 2010, which *The Washington Post* called "the most significant tax bill in nearly a decade",[140] passed with bipartisan majorities in both houses of Congress and was signed into law by Obama on December 17, 2010.[138]

Ethics

Lobbying reform

Early in his presidential campaign, Obama stated that "they [lobbyists] won't find a job in my White House", but softened his stance later in the campaign.[141] On January 21, 2009, Obama issued an executive order for all future appointees to his administration, which stated, no appointee who was a registered lobbyist within the two years before his appointment could participate on matters in which he lobbied for a period of two years after the date of appointment.[41] Three formal waivers were initially issued in early 2009, out of 800 executive appointments:[142] [143] to William J. Lynn III, a lobbyist for Raytheon, to hold the position of Deputy Secretary of Defense;[35] to Jocelyn Frye, former general counsel at the National Partnership for Women and Families, to serve as Director of

Policy and Projects in the Office of the First Lady; and to Cecilia Muñoz, former senior vice president for the National Council of La Raza,[142] to serve as Director of Intergovernmental Affairs in the Executive Office of the President.[143] As of March 21, 2009, at least thirty officials appointed by Obama had been lobbyists in the past five years.[141] Ten additional waivers were announced in September 2009.[144]

Not all recent former lobbyists require waivers; those without waivers write letters of recusal stating issues from which they must refrain because of their previous jobs.[142] *USA Today* reported that 21 members of the Obama administration have at some time been registered as federal lobbyists, although most have not within the previous two years.[145] Lobbyists in the administration include William Corr, an anti-tobacco lobbyist, as Deputy Secretary of Health and Human Services[146] and Tom Vilsack, who lobbied in 2007, for a national teachers union, as Secretary of Agriculture.[145] Also, the Secretary of Labor nominee, Hilda Solis, formerly served as a board member of American Rights at Work, which lobbied Congress on two bills Solis co-sponsored,[147] and Mark Patterson, Treasury Secretary Timothy Geithner's chief of staff, is a former lobbyist for Goldman Sachs.[145]

The Citizens for Responsibility and Ethics in Washington have criticized the administration, claiming that Obama is retreating from his own ethics rules barring lobbyists from working on the issues about which they lobbied during the previous two years by issuing waivers. According to Melanie Sloan, the group's executive director, "It makes it appear that they are saying one thing and doing another."[145]

Transparency

The Obama administration has said that all executive orders, non-emergency legislation, and proclamations will be posted to the official White House website, whitehouse.gov, allowing the public to review and comment for five days before the President signs the legislation.[148] The pledge was twice broken during Obama's first month in office when he signed SCHIP legislation and the Lilly Ledbetter Fair Pay Act with less than the full five days of "sunlight before signing". The administration has said that they are still "working through implementation procedures and some initial issues with the congressional calendar".[149] [150]

Obama presents his first weekly address as President of the United States, discussing the American Recovery and Reinvestment Act of 2009.

During his first week in office, Obama announced plans to post a video address each week on the site,[45] and on YouTube,[46] informing the public of government actions each week. During his speech at the 2008 Democratic National Convention, Obama stated, "I will also go through the federal budget, line by line, eliminating programs that no longer work and making the ones we do need work better and cost less - because we cannot meet twenty-first century challenges with a twentieth century bureaucracy."[151]

On January 21, 2009, by executive order, Obama revoked Executive Order 13233, which had limited access to the records of former United States Presidents.[152] Obama issued instructions to all agencies and departments in his administration to "adopt a presumption in favor" of Freedom of Information Act requests.[153] In April 2009, the United States Department of Justice released four legal memos from the Bush administration to comply voluntarily with a Freedom of Information Act lawsuit filed by the American Civil Liberties Union.[154] The memos were written by John Yoo[155] and signed by Jay Bybee and Steven Bradbury, then Principal Assistant Attorneys General to the Department of Justice, and addressed to John A. Rizzo, general counsel of the Central Intelligence Agency.[156] The memos describe in detail controversial interrogation methods the CIA used on prisoners suspected of terrorism.[157] [158] Obama became personally involved in the decision to release the memos, which was opposed by former CIA directors Michael Hayden, Porter Goss, George Tenet and John Deutch.[156] Former Vice President Dick Cheney criticized Obama for not releasing more memos; Cheney claimed that unreleased memos detail successes of CIA interrogations.[159]

The American Recovery and Reinvestment Act requires all recipients of the funds provided by the act to publish a plan for using the funds, along with purpose, cost, rationale, net job creation, and contact information about the plan to a website Recovery.gov [160] so that the public can review and comment. Inspectors General from each department or executive agency will then review, as appropriate, any concerns raised by the public. Any findings of an Inspector General must be relayed immediately to the head of each department and published on Recovery.gov.[161]

On June 16, 2009, Citizens for Responsibility and Ethics in Washington (CREW) filed a lawsuit against the Obama administration in order to get information about the visits of coal company executives. Anne Weismann, the chief counsel for CREW, stated "The Obama administration has now taken exactly the same position as the Bush administration... I don't see how you can keep people from knowing who visits the White House and adhere to a policy of openness and transparency." On June 16, MSNBC reported that its more comprehensive request for visitor logs since Obama's January 20 inauguration had been denied.[162] The administration announced that White House visitor logs will be made available to the public on an ongoing basis, with certain limitations, for visits occurring after September 15, 2009.[163] Beginning on January 29, 2010, the White House did begin to release the names of its visitor records.[164] Since that time, names of visitors (which includes not only tourists, but also names of union leaders, Wall Street executives, lobbyists, party chairs, philanthropists and celebrities), have been released. The names are released in huge batches up to 75,000 names at a time.[165] Names are released 90–120 days after having visited the White House. The complete list of names is available online by accessing the official White House website.[166]

Obama stated during the 2008 Presidential campaign that he would have negotiations for health care reform televised on C-SPAN, citing transparency as being the leverage needed to ensure that people stay involved in the process taking place in Washington. This did not fully happen and Politifact gives President Obama a "Promise Broken" rating on this issue.[167] After White House press secretary Robert Gibbs initially avoided addressing the issue,[168] President Obama himself acknowledged that he met with Democratic leaders behind closed doors to discuss how best to garner enough votes in order to merge the two (House and Senate) passed versions of the health care bill. Doing this violated the letter of the pledge, although Obama maintains that negotiations in several congressional committees were open, televised hearings. Obama also cited an independent ethics watchdog group describe his administration as the most transparent in recent history.[169]

The Obama administration has been characterized[170] as much more aggressive than the Bush and other previous administrations in their response to whistleblowing and leaks to the press. Three people have been prosecuted under the rarely used Espionage Act of 1917. They include Thomas Andrews Drake, a former National Security Agency (NSA) employee who was critical of the NSA's Trailblazer Project,[171] [172] [173] Stephen Jin-Woo Kim, a State Department contractor who allegedly had a conversation about North Korea with James Rosen of Fox News,[174] [175] and Jeffrey Sterling, who allegedly was a source for James Risen's book State of War. Risen has also been subpoenaed to reveal his sources, another rare action by the government.[176]

Foreign policy

The first meeting between Dmitry Medvedev and Barack Obama before the G20 summit in London on April 1, 2009.

In his inaugural address, Obama suggested that he plans to begin the process of withdrawing from Iraq and continuing to focus on the war in Afghanistan. He also mentioned lessening the nuclear threat through "working tirelessly with old friends and former foes". He spoke about America's determination to combat terrorism, proclaiming America's spirit is "stronger and cannot be broken — you cannot outlast us, and we will defeat you." To the Muslim world, Obama extended an invite to "a new way forward, based on mutual interest and mutual respect". He also said the USA would "extend a hand" to those "who cling to power through corruption and deceit" if they "are willing to unclench" their fists.[177] Shortly after his inauguration President Obama first called President Abbas of the Palestinian National Authority (PNA). Calls were also made to President Mubarak of Egypt, Prime Minister Olmert of Israel and King Abdullah of Jordan.[178] Obama and Secretary of State Hillary Clinton named George Mitchell as Special Envoy for Middle East peace and Richard Holbrooke as special representative to Pakistan and Afghanistan on January 23, 2009.[179] At the same time, Obama called on Israel to open the borders of Gaza, detailing early plans on his administration's peace plans for the Israeli–Palestinian conflict.[180]

President Obama and First Lady Michelle Obama are welcomed by Queen Elizabeth II to Buckingham Palace in London, April 1, 2009.

On February 18, 2009, Obama announced that the U.S. military presence in Afghanistan would be bolstered by 17,000 new troops by summer.[181] The announcement followed the recommendation of several experts including Defense Secretary Robert Gates that additional troops be deployed to the war-torn nation.[182] [183]

Obama declared his plan for ending the Iraq War on February 27, 2009, in a speech at Camp Lejeune, North Carolina, before an audience of Marines stationed there. According to the president, combat troops will be withdrawn from Iraq by August 2010, leaving a contingent of up to 50,000 servicemen and servicewomen to continue training, advisory, and counterterrorism operations until as late as the end of 2011.[184] [185]

Other characteristics of the Obama administration on foreign policy include a tough stance on tax havens,[186] continuing military operation in Pakistan,[187] and avowed focus on diplomacy to prevent nuclear proliferation in Iran[188] and North Korea.[189]

On April 1, 2009, Obama and China's President, Hu Jintao, announced the establishment of the U.S.-China Strategic and Economic Dialogue and agreed to work together to build a positive, cooperative, and comprehensive U.S.-China relationship for the 21st century.[190]

In that same month, Obama requested that Congress approve $83.4 billion of supplemental military funding, mostly for the war in Iraq and to increase troop levels in Afghanistan. The request also includes $2.2 billion to increase the size of the US military, $350 million to upgrade security along the US-Mexico border, and $400 million in counterinsurgency aid for Pakistan.[191]

In May 2009, it was reported that Obama plans to expand the military by 20,000 employees.[192]

On June 4, 2009, Obama delivered a speech at Cairo University in Egypt.[193] The wide ranging speech called for a "new beginning" in relations between the Islamic world and the United States.[194] [195] The speech received both praise and criticism from leaders in the region.[196] [197] [198] [199] [200] In March 2010, Secretary of State Clinton criticized the Israeli government for approving expansion of settlements in East Jerusalem.[201]

On April 8, 2010, Obama and Russian President Dmitry Medvedev signed the latest Strategic Arms Reduction Treaty (START), a "major" nuclear arms control agreement that reduces the nuclear weapons stockpiles of both countries.[202]

In March 2011, international reaction to Muammar Gaddafi's military crackdown on rebel forces and civilians in Libya culminated in a United Nations resolution to enforce a no fly zone in Libya. Obama authorized U.S. forces to participate in international air attacks on Libyan air defenses using Tomahawk cruise missiles to establish the protective zone.[203]

Guantánamo Bay detention camp

On his first day in office, Obama requested a 120-day suspension of all trials for alleged terrorists held at the Guantánamo Bay detention camp, so the new administration could "review the military commissions process, generally, and the cases pending before military commissions as of 2011, specifically".[204] Another order established a task force to lead a review of detention policies, procedures and individual cases. Obama addressed the State Department that "the United States will not torture" and drafted an executive order to close Guantánamo within a year.[205] On January 22, 2009, Obama signed an executive order ensuring safe, lawful, and humane treatment of individuals detained in armed conflicts. This order restricts interrogators to methods listed and authorized by an Army Field Manual.[206] A detainee released since Obama took office claimed in an interview with Agence France-Presse that conditions at Guantánamo had worsened, stating guards wanted to "take their last revenge" before the facility is closed.[207] On March 13, 2009, the administration announced that it would no longer refer to prisoners at Guantánamo Bay as enemy combatants, but it also asserted that the president has the authority to detain terrorism suspects there without criminal charges.[208]

The case review of detainee files by administration officials and prosecutors was made more difficult than expected as the Bush administration had failed to establish a coherent repository of the evidence and intelligence on each prisoner. By September 2009, prosecutors recommended to the Justice Department which detainees are eligible for trial, and the Justice Department and the Pentagon worked together to determine which of several now-scheduled trials will go forward in military tribunals and which in civilian courts. While 216 international terrorists are already held in maximum security prisons in the U.S., Congress was denying the administration funds to shut down the camp and adapt existing facilities elsewhere, arguing that the decision was "too dangerous to rush".[209] In November, Obama stated that the U.S. would miss the January 2010 date for closing the Guantánamo Bay prison as he had ordered, acknowledging that he "knew this was going to be hard". Obama did not set a specific new deadline for closing the camp, citing that the delay was due to politics and lack of congressional cooperation.[210] The state of Illinois has offered to sell to the federal government the Thomson Correctional Center, a new but largely unused prison, for the purpose of housing detainees. Federal officials testified at a December 23 hearing that if the state commission approves the sale for that purpose, it could take more than six months to ready the facility.[211]

Killing of Osama bin Laden

Obama, sitting next to Biden, with the U.S. national security team gathered in the Situation Room to monitor the progress of Operation Neptune Spear.

Starting with information received in July 2010, intelligence developed by the CIA over the next several months determined what they believed to be the location of Osama bin Laden in a large compound in Abbottabad, Pakistan, a suburban area 35 miles from Islamabad.[212] CIA head Leon Panetta reported this intelligence to Obama in March 2011.[212] Meeting with his national security advisers over the course of the next six weeks, Obama rejected a plan to bomb the compound, and authorized a "surgical raid" to be conducted by United States Navy SEALs.[212] The operation took place on May 1, 2011, resulting in the death of bin Laden and the seizure of papers and computer drives and disks from the compound.[213] [214] Bin Laden's body was identified through DNA testing,[215] and buried at sea several hours later.[216] Within minutes of Obama's announcement from Washington, DC, late in the evening on May 1, there were spontaneous celebrations around the country as crowds gathered outside the White House, and at New York City's Ground Zero and Times Square.[213] [217] Reaction to the announcement was positive across party lines, including from predecessors George W. Bush and Bill Clinton,[218] and from many countries around the world.[219]

Overseas Contingency Operation

Obama discontinued use of the term "War on Terror" and instead uses the term "Overseas Contingency Operation". However, Obama has stated that the U.S. is at war with Al-Qaeda, saying "Our nation is at war, against a far-reaching network of violence and hatred."[220]

In April 2010, the Obama administration authorized the targeted killing of the radical Muslim cleric and American citizen Anwar al-Awlaki, who was believed to have shifted from encouraging attacks on the United States to directly participating in them.[221] The Justice Department legal memorandum authorizing the strike, which asserted that Fifth Amendment due process rights "could be satisfied by internal deliberations in the executive branch", has not been released to the public.[222]

Gun control

During the presidential campaign, Obama announced that he favors measures that respect Second Amendment rights, while at the same time keeping guns away from children and criminals.[223] [224] On February 25, 2009, Attorney General Eric Holder announced that the Obama administration would seek a new assault weapons ban across the United States, saying that it would have a positive impact on the drug-related violence in Mexico.[225] After the statement drew criticism from the NRA and some House Democrats, the Administration reportedly ordered the Justice Department to end public discussion of the issue.[226] Obama has signed into law two bills containing amendments reducing restrictions on gun owners, one which permits guns to be transported in checked baggage on Amtrak trains[227] and another which allows carrying loaded firearms in national parks located in states allowing concealed carry.[228] [229]

Science and technology

Cybersecurity

President Obama during a call to the crew aboard the ISS.

Obama initiated a 60-day review of cybersecurity[230] by Melissa Hathaway, a consultant for Booz Allen Hamilton, appointed Acting Senior Director for Cyberspace.[231] [232]

The New York Times reported in 2009, that the NSA is intercepting communications of American citizens including a Congressman, although the Justice Department believed that the NSA had corrected its errors.[233] United States Attorney General Eric Holder resumed the wiretapping according to his understanding of the Foreign Intelligence Surveillance Act of 1978 Amendments Act of 2008 that Congress passed in July 2008, but without explaining what had occurred.[234]

Environment

On January 27, 2009, Obama issued two presidential memoranda concerning energy policy. One directed the Department of Transportation to raise fuel efficiency standards incrementally to 35 miles per US gallon (**unknown operator: u'strong'** km/L) by 2020, and the other directed the Environmental Protection Agency to allow individual states to set stricter tailpipe emissions regulations than the federal standard.[42] [235]

The American Recovery and Reinvestment Act of 2009 provides $54 billion in funds to double domestic renewable energy production, renovate federal buildings making them more energy-efficient, improve the nation's electricity grid, repair public housing, and weatherize modest-income homes.[236]

On February 10, 2009, Obama overturned a Bush administration policy that had opened up a five-year period of offshore drilling for oil and gas near both the Atlantic and Pacific coasts. Interior Secretary Ken Salazar has been quoted as saying, "To establish an orderly process that allows us to make wise decisions based on sound information, we need to set aside" the plan "and create our own timeline".[237]

On May 19, 2009, Obama announced a plan to increase the Corporate Average Fuel Economy national standards for gasoline mileage, by creating a single new national standard that will create a car and light truck fleet in the United States that is almost 40 percent cleaner and more fuel-efficient by 2016, than it is today, with an average of 35.5 miles per gallon.[238] Environmental advocates and industry officials welcomed the new program, but for different reasons. Environmentalists called it a long-overdue tightening of emissions and fuel economy standards after decades of government delay and industry opposition. Auto industry officials said it would provide the single national efficiency standard they have long desired, a reasonable timetable to meet it and the certainty they need to proceed with product development plans.[238]

On March 30, 2010, Obama partially reinstated Bush administration proposals to open certain offshore areas along the Atlantic coastline, the eastern Gulf of Mexico and the north coast of Alaska to oil and natural gas drilling. The proposals had earlier been set aside by President Obama after they were challenged in court on environmental grounds.[239]

On May 27, 2010, Obama extended a moratorium on offshore drilling permits after the April 20, 2010 Deepwater Horizon oil spill which is considered to be the worst oil spill in U.S. history.[240] [241] Although BP took responsibility for the disaster and its ongoing after effects, Obama began a federal investigation along with forming a bipartisan commission to review the incident and methods to avoid it in the future.[242] [243] [244] Obama visited the Gulf Coast on May 2 and May 28 and expressed his frustration on the June 8 *NBC Today Show*, by saying "I don't sit around just talking to experts because this is a college seminar. We talk to these folks because they potentially have the best answers, so I know whose ass to kick."[245] Obama's response to the disaster drew confusion and criticism within segments of the media and public.[246]

NASA

President Obama speaks at Kennedy Space Center, April 15, 2010.

Obama set up the Augustine panel to review the Constellation program in 2009, and announced in February 2010, that he was cutting the program from the 2011 United States federal budget, describing it as "over budget, behind schedule, and lacking in innovation."[247] [248] [249] [250] [251] After the decision drew criticism in the United States, a new "Flexible path to Mars" plan was unveiled at a space conference in April 2010.[252] [253] [254] It included new technology programs, increased R&D spending, a focus on the International Space Station and contracting out flying crew to space to commercial providers.[255] The new plan also increased NASA's 2011 budget to $19 billion from $18.3 billion in 2010.[252]

In July 2009, Obama appointed Charles Bolden, a former astronaut, to be administrator of NASA.[256]

Stem cell research

On March 9, 2009, Obama repealed a Bush-era policy that prevented federal tax dollars from being used to fund research on new lines of embryonic stem cells. Such research has been a matter of debate between those who emphasize the therapeutic potential of such research and those who suggest that elements of this research breach ethical limitations. Obama stated that "In recent years, when it comes to stem cell research, rather than furthering discovery, our government has forced what I believe is a false choice between sound science and moral values...In this case, I believe the two are not inconsistent. As a person of faith, I believe we are called to care for each other and work to ease human suffering. I believe we have been given the capacity and will to pursue this research — and the humanity and conscience to do so responsibly."[257]

Social policy

On January 23, 2009, Obama rescinded the Mexico City Policy, a measure from the Reagan and Bush eras that required any non-governmental organization receiving U.S. Government funding to refrain from performing or promoting abortion services in other countries.[258]

On June 17, 2009, Obama authorized the extension of some benefits (but not health insurance or pension benefits) to same-sex partners of federal employees.[259] Obama has chosen to leave larger changes, such as the repeal of Don't ask, don't tell and the Defense of Marriage Act, to Congress.[260] [261]

On October 19, 2009, the U.S. Department of Justice issued a directive to federal prosecutors in states with medical marijuana laws not to investigate or prosecute cases of marijuana use or production done in compliance with those laws.[262]

On December 16, 2009, President Obama signed the Consolidated Appropriations Act, 2010, which repealed a 21-year-old ban on federal funding of needle exchange programs.[263]

On December 22, 2010, Obama signed the Don't Ask, Don't Tell Repeal Act of 2010, a bill that provides for repeal of the Don't ask, don't tell policy of 1993, that has prevented gay and lesbian people from serving openly in the United States Armed Forces.[264] Repealing "Don't ask, don't tell" had been a key campaign promise that Obama had made during the 2008 presidential campaign.[265] [266]

Health care reform

Once the stimulus bill was enacted, health care reform became Obama's top domestic priority. On July 14, 2009, House Democratic leaders introduced a 1,000 page plan for overhauling the US health care system, which Obama wanted Congress to approve by the end of the year.[267]

The U.S. Congressional Budget Office (CBO) estimated the ten-year cost to the federal government of the major insurance-related provisions of the bill at approximately $1.0 trillion.[268] In mid-July 2009, Douglas Elmendorf, director of the CBO, testified that the proposals under consideration would significantly increase federal spending and did not include the "fundamental changes" needed to control the rapid growth in health care spending.[269] [270] However after reviewing the final version of the bill introduced after 14 months of debate the CBO estimated that it would reduce federal budget deficits by $143 billion over 10 years and by more than a trillion in the next decade.[271]

After much public debate during the Congressional summer recess of 2009, Obama delivered a speech to a joint session of Congress on September 9 where he addressed concerns over his administration's proposals.[272] In March 2010, Obama gave several speeches across the country to argue for the passage of health care reform.[273] [274] On March 21, 2010, after Obama announced an executive order reinforcing the current law against spending federal funds for elective abortion services, the House, by a vote of 219 to 212, passed the version of the bill previously passed on December 24, 2009, by a 60-vote supermajority in the Senate. The bill, which includes over 200 Republican amendments, was passed without a single Republican vote. On March 23, 2010, President Obama signed the bill into law. Immediately following the bill's passage, the House voted in favor of a reconciliation measure to make significant changes and corrections to the Patient Protection and Affordable Care Act, which was passed by both houses with two minor alterations on March 25, 2010, and signed into law on March 30, 2010.[275] [276]

Education reform

On March 30, 2010, Obama signed the Health Care and Education Reconciliation Act of 2010, which ended the role of private banks in lending out federally-insured student loans.[277] By directly lending to students, the government is projected to save taxpayers $68 billion dollars over the next several years.[278] Federally-insured student loans will instead be distributed by the Department of Education.[279] The law also increased the amount of Pell Grant awards given each year, doubling its current funding.[280] [281] Starting in 2014, the law permits borrowers to cap the amount they spend on student loans each year to ten percent of their discretionary income and have their balance forgiven if they have faithfully paid the balance of their loan over 20 years.[280] [282] Additionally, the law seeks to make it easier for parents to qualify for Grad PLUS loans, and spends billions on poor and minority schools and $2 billion for community colleges.[278] [279]

Gates arrest controversy

Further information: Henry Louis Gates arrest controversy

On July 16, 2009, prominent African American Harvard University professor Henry Louis Gates, Jr., was arrested at his Cambridge, Massachusetts home by a local white police officer, Cambridge Police Sgt. James Crowley, for disorderly conduct. Gates, who was locked out of his house, had attempted to break into his own property, thus causing the initial alarm from a neighbor who called 9-1-1. The incident sparked national controversy over whether Gates's civil rights had been violated by Crowley. On July 21, the Cambridge Police Department dropped charges against Gates. On July 22, President Barack Obama, commented on the incident over national and international television, criticized the arrest, and stated the police acted "stupidly" in handling the incident. National law enforcement organizations and members objected to Obama's comments and criticized his handling of the issue.

"Beer Summit" at the White House, July 30, 2009.
From left to right: Gates, Crowley, and President Obama.
Photo taken before Vice President Biden's arrival.

In the aftermath, Obama stated that he regretted his comments exacerbating the situation, and hoped that the situation could become a "teachable moment". To reduce tensions, on July 24, Obama invited both parties to the White House to discuss the issue over beers, and on July 30, Obama and Vice President Joe Biden joined Crowley and Gates in a private, cordial meeting in a courtyard near the White House Rose Garden; this became known colloquially as the "Beer Summit".[283]

Wall Street reform

On July 21, 2010, Obama signed Dodd–Frank Wall Street Reform and Consumer Protection Act, considered to be the largest financial system overhaul since the New Deal. The law recognizes complex financial derivatives and makes rules to protect consumers from unfair practices in loans and credit cards by establishing a new consumer protection agency. At the signing ceremony in the Ronald Reagan Building in Washington D.C. Obama proclaimed, "There will be no more taxpayer-funded bailouts. Period." Obama also mentioned that "These reforms represent the strongest consumer financial protections in history." At the ceremony were Sen. Christopher Dodd, D-Conn., and Rep. Barney Frank, D-Mass., the two committee chairmen who sponsored the bill.[284]

2010 midterm elections

Due in large part to voter frustration over high unemployment and a stalled economy, Republicans won control of the House in the 2010 midterm elections. On reflection after the election, Obama blamed himself, in part, for the many Democrats who went down to defeat knowing that they had risked their careers to support his agenda of economic stimulus legislation and a landmark health care bill.[285] Democrats narrowly retained the Senate majority and will continue to control it through the 112th Congress.[286]

Obama called the elections "humbling" and a "shellacking".[287] He said that the results came because not enough Americans had felt the effects of the economic recovery.[288]

2012 re-election campaign

On April 4, 2011, President Obama announced that he would seek re-election in the 2012 presidential election. The campaign will be based in Chicago and is being run by many former members of the White House staff and members of the successful 2008 campaign.[289]

References

[1] "Obama's birth certificate: Final chapter" (http://www.politifact.com/truth-o-meter/article/2008/jun/27/obamas-birth-certificate-part-ii/). Politifact. June 27, 2008. . Retrieved December 12, 2008.

[2] "Obama's church choice likely to be scrutinized" (http://www.msnbc.msn.com/id/27775757/). Associated Press. msnbc.com. November 17, 2008. . Retrieved January 20, 2009.

[3] "Barack Obama, long time UCC member, inaugurated forty-fourth U.S. President" (http://www.ucc.org/news/obama-inauguration.html) (Press release). United Church of Christ. January 20, 2009. . Retrieved January 21, 2009. "Barack Obama, who spent more than 20 years as a UCC member, is the forty-fourth President of the United States."

[4] An Associated Press wire story on Obama's resignation from Trinity United Church of Christ in the course of the Jeremiah Wright controversy stated that he had, in doing so, disaffiliated himself with the UCC. (See "Obama's church choice likely to be scrutinized" (http://www.msnbc.msn.com/id/27775757/). Associated Press. msnbc.com. November 17, 2008. . Retrieved January 20, 2009.)

[5] http://www.whitehouse.gov/

[6] Zeleny, Jeff; Baker, Peter (November 6, 2008). "Rahm Emanuel Accepts Post as White House Chief of Staff" (http://www.nytimes.com/2008/11/06/us/politics/07elect.html). *The New York Times*. . Retrieved December 6, 2008.

[7] "Geithner, Summers among key economic team members announced today" (http://change.gov/newsroom/entry/geithner_summers_among_key_economic_team_members_announced_today/) (Press release). Office of the President-elect. November 24, 2008. . Retrieved December 6, 2008.

[8] Baker, Peter (November 25, 2008). "Defense Secretary Said to Be Staying On" (http://thecaucus.blogs.nytimes.com/2008/11/25/defense-secretary-said-to-be-staying-on/). *The New York Times*. . Retrieved December 6, 2008.

[9] Baker, Peter (November 30, 2008). "Obama's Choice for U.N. Is Advocate of Strong Action Against Mass Killings" (http://www.nytimes.com/2008/12/01/us/politics/01rice.html). *The New York Times*. . Retrieved December 6, 2008.

[10] "Official Obama-Biden Transition Website" (http://change.gov/content/home). Office of the President-elect. . Retrieved January 26, 2009.

[11] "Obama Transition announces rules for lobbyists in Transition" (http://change.gov/newsroom/entry/obama_transition_announces_rules_for_lobbyists_in_transition) (Press release). Office of the President-elect. November 11, 2008. . Retrieved January 26, 2009.

[12] "Citizen's Briefing Book" (http://citizensbriefingbook.change.gov/). Office of the President-elect. . Retrieved January 26, 2009.

[13] Marisa Taylor (January 20, 2009). "Change Has Come To WhiteHouse.Gov" (http://blogs.wsj.com/digits/2009/01/20/change-has-come-to-whitehousegov/). *The Wall Street Journal*. . Retrieved February 13, 2009.

[14] United States Constitution. "20th Amendment to the United States Constitution" (http://www.law.cornell.edu/constitution/constitution.amendmentxx.html). . Retrieved January 21, 2009.

[15] "Obama Signs First Presidential Proclamation" (http://politicalticker.blogs.cnn.com/2009/01/20/obama-signs-first-presidential-proclamation/). *CNN*. January 20, 2009. . Retrieved January 20, 2009.

[16] Kelley, Matt (January 19, 2009). "Obama to be sworn in on 'Lincoln Bible'" (http://www.usatoday.com/news/washington/2009-01-19-Lincoln-bible_N.htm). *USA Today*. . Retrieved January 26, 2009.

[17] "Chief justice fumbles oath" (http://www.washingtontimes.com/news/2009/jan/20/obama-roberts-fumble-oath/). *The Washington Times*. January 20, 2009. . Retrieved January 25, 2009.

[18] "Obama takes presidential oath again after stumble" (http://www.chinadaily.com.cn/world/2009-01/22/content_7421457.htm). China Daily. January 22, 2009. . Retrieved March 10, 2010.

[19] Chernus, Ira (December 16, 2008). "The First Hundred Days or the Last Hundred Days?" (http://www.laprogressive.com/2008/12/16/the-first-hundred-days-or-the-last-hundred-days/). *The LA Progressive*. . Retrieved January 18, 2009.

[20] Reid, Tim (November 1, 2008). "Barack Obama lays plans to deaden expectation after election victory" (http://www.timesonline.co.uk/tol/news/world/us_and_americas/us_elections/article5051118.ece). London: The Times. . Retrieved January 18, 2009.

[21] Dorning, Mike (May 28, 2008). "Obama's priorities for first 100 days" (http://www.swamppolitics.com/news/politics/blog/2008/05/obamas_priorities_for_first_10.html). *Chicago Tribune*. . Retrieved January 18, 2009.

[22] Examples of "100 days" – portals:

- "Obama's First 100 Days" (http://www.huffingtonpost.com/news/100-days). *The Huffington Post*. . Retrieved January 21, 2009.
- "Special Reports - The First 100 Days of the 44th President Barack Obama" (http://www.cnn.com/SPECIALS/2009/44.president/first.100.days/). CNN. . Retrieved 14 May 2012.
- "The 44th President 100 Days" (http://www.cbsnews.com/sections/politics/100days/main503723.shtml). CBS News. . Retrieved January 21, 2009.

[23] Hall, Ashley (January 21, 2009). "Domestic issues to dominate Obama's first 100 days in office" (http://www.abc.net.au/pm/content/2008/s2471452.htm). ABC Online, Australia. . Retrieved January 21, 2009.

[24] Naughton, Philippe (January 21, 2009). "Barack Obama calls halt to Guantanamo trials" (http://www.timesonline.co.uk/tol/news/world/us_and_americas/us_elections/article5558726.ece). London: The Times. . Retrieved January 21, 2009.

[25] Sahadi, Jeanne (January 20, 2009). "First 100 days: Obama's burden" (http://money.cnn.com/2009/01/20/news/economy/obama_first100days/?postversion=2009012012). CNN. . Retrieved January 21, 2009.

[26] West, Larry (January 20, 2009). "Obama's First 100 Days: An Environmental Agenda for Obama's First 100 Days" (http://environment.about.com/b/2009/01/20/obamas-first-100-days-an-environmental-agenda-for-obamas-first-100-days.htm). About.com. . Retrieved January 21, 2009.

[27] Murray, Lisa (January 22, 2009). "Jakarta schoolchildren cheer for the rise of a former student" (http://www.smh.com.au/news/world/jakarta-schoolchildren-cheer-for-the-rise-of-a-former-student/2009/01/21/1232471395138.html). *The Sydney Morning Herald*. . Retrieved January 21, 2009.

[28] Smith, Jean Edward (January 16, 2009). "How F.D.R. Made the Presidency Matter" (http://100days.blogs.nytimes.com/2009/01/16/how-fdr-made-the-presidency-matter/). *The New York Times* (The New York Times Company). . Retrieved January 29, 2009.

[29] Reeves, Richard (January 19, 2009). "Kennedy's Words, Obama's Challenge" (http://100days.blogs.nytimes.com/2009/01/19/kennedys-words-obamas-challenge/). *The New York Times* (The New York Times Company). . Retrieved January 29, 2009.

[30] Dallek, Richard (January 23, 2009). "L.B.J., Obama and Reassuring a Worried Nation" (http://100days.blogs.nytimes.com/2009/01/23/lbj-obama-and-reassuring-a-worried-nation/). *The New York Times* (The New York Times Company). . Retrieved January 29, 2009.

[31] Smith, Jean Edward (January 27, 2009). "Obama's Reagan Transformation?" (http://100days.blogs.nytimes.com/2009/01/27/obamas-reagan-transformation/). *The New York Times* (The New York Times Company). . Retrieved January 29, 2009.

[32] "Obama halts all regulations pending review" (http://www.msnbc.msn.com/id/28758810/). Associated Press. msnbc.com. January 20, 2009. . Retrieved January 21, 2009.

[33] "Obama's first day: Pay freeze, lobbying rules" (http://www.msnbc.msn.com/id/28767687/). msnbc.com. January 21, 2009. . Retrieved January 21, 2009.

[34] Memmott, Mark (January 21, 2009). "Revelry Over, Obama freezing pay of top staff; signs ethics rules" (http://content.usatoday.com/communities/theoval/post/2009/01/61712448/1). *USA Today*. . Retrieved January 21, 2009.

[35] "Obama breaks his own rule" (http://www.cnn.com/2009/POLITICS/01/23/campbell.brown.lobbyists/index.html). CNN. January 23, 2009. . Retrieved January 23, 2009.

[36] "Obama grants two more lobbyists waivers" (http://thehill.com/business-a-lobbying/3960-obama-grants-two-more-lobbyists-waivers). *The Hill*. March 10, 2009. . Retrieved 2010-12-18.

[37] Mazzetti, Mark; Glaberson, William (January 21, 2009). "Obama Issues Directive to Shut Down Guantánamo" (http://www.nytimes.com/2009/01/22/us/politics/22gitmo.html). *The New York Times*. . Retrieved May 22, 2010.

[38] "Closure Of Guantanamo Detention Facilities" (http://www.whitehouse.gov/the_press_office/ClosureOfGuantanamoDetentionFacilities/). The White House. January 22, 2009. . Retrieved January 27, 2009.

[39] "Obama signs order to close Guantanamo in a year" (http://www.washingtontimes.com/news/2009/jan/22/obama-signs-order-close-guantanamo-year/). *The Washington Times*. January 22, 2009. . Retrieved March 19, 2010.

[40] Obama Reverses Key Bush Security Policies (http://www.nytimes.com/2009/01/23/us/politics/23obama.html), *The New York Times*, January 22, 2009

[41] "Ethics Commitments by Executive Branch Personnel" (http://www.whitehouse.gov/the_press_office/ExecutiveOrder-EthicsCommitments). The White House. . Retrieved January 22, 2009.

[42] "From Peril to Progress" (http://www.whitehouse.gov/blog_post/Fromperiltoprogress/). The White House. January 26, 2009. . Retrieved January 26, 2009.

[43] "Obama ends funding ban for abortion groups abroad" (http://www.reuters.com/article/newsOne/idUSTRE50M3PQ20090123). *Reuters*. January 23, 2009. . Retrieved January 23, 2009.

[44] Obama reverses Bush abortion-funds policy (http://www.nytimes.com/aponline/2009/01/23/washington/AP-Obama-Abortion-Ban.html). January 23, 2009.

[45] "President Barack Obama's Weekly Video Address" (http://web.archive.org/web/20090122233027/http://www.whitehouse.gov/weekly_address/). The White House. January 22, 2009. Archived from the original (http://www.whitehouse.gov/weekly_address/) on January 22, 2009. . Retrieved September 8, 2011.

[46] "Official White House YouTube Channel" (http://www.youtube.com/user/whitehouse). *YouTube*. YouTube, LLC. January 24, 2009. . Retrieved January 26, 2009.

[47] Baker, Peter (January 24, 2009). "In Effort to Build Support, Obama Details Stimulus Plan" (http://www.nytimes.com/2009/01/25/us/politics/25obama.html). *The New York Times* (The New York Times Company). . Retrieved January 26, 2009.

[48] Bleeker, Andrew (November 15, 2008). "Your Weekly Address from the President-elect" (http://change.gov/newsroom/entry/your_weekly_address_from_the_president_elect/). *Change.gov*. The Office of the President-elect. . Retrieved January 26, 2009.

[49] Vargas, Jose Antonio (November 14, 2008). "The YouTube Presidency" (http://voices.washingtonpost.com/44/2008/11/14/the_youtube_presidency.html). *The Washington Post* (The Washington Post Company). . Retrieved January 26, 2009.

[50] Rutenberg, Jim and Adam Nagourney (January 25, 2009). "Melding Obama's Web to a YouTube Presidency" (http://www.nytimes.com/2009/01/26/us/politics/26grassroots.html). *The New York Times* (The New York Times Company). . Retrieved January 26, 2009.

[51] "A Wonderful Day" (http://www.whitehouse.gov/blog/2009/01/29/a-wonderful-day). The White House. January 29, 2009. . Retrieved January 29, 2009.

[52] "CHIP" (http://www.whitehouse.gov/blog_post/chip/). The White House.

[53] http://clerk.house.gov/evs/2009/roll070.xml

[54] "Congress Readies Final Vote On $790B Stimulus Bill" (http://www.wusa9.com/news/watercooler/story.aspx?storyid=81491). WUSA. . Retrieved 14 May 2012.

[55] Associated Press (13 February 2009). "" (http://www.columbiatribune.com/news/2009/feb/13/congress-readies-final-vote-on-stimulus/). *Columbia Daily Tribune*. . Retrieved 14 May 2012.

[56] Sahadi, Jeanne (2009-01-27). "Stimulus with interest: $1.2 trillion" (http://money.cnn.com/2009/01/27/news/economy/stimulus_spending). Money.cnn.com. . Retrieved 2012-04-14.

[57] "Stimulus: Now for the hard part" (http://money.cnn.com/2009/02/17/news/economy/obama_stimulus_meas_success/index.htm) CNN. February 17, 2009.

[58] Stolberg, Sheryl (March 9, 2009). "Obama Lifts Bush's Strict Limits on Stem Cell Research" (http://www.nytimes.com/2009/03/10/us/politics/10stem.html). *The New York Times*. . Retrieved March 17, 2009.

[59] Savage, Charlie (March 9, 2009). "Obama Looks to Limit Impact of Tactic Bush Used to Sidestep New Laws" (http://www.nytimes.com/2009/03/10/us/politics/10signing.html). *The New York Times*. . Retrieved March 17, 2009.

[60] Obama, Barack (March 9, 2009). "Subject: Presidential Signing Statements" (http://graphics.nytimes.com/packages/pdf/politics/2009signingstm.mem.final.rel.pdf). The White House. . Retrieved March 17, 2009.

[61] Savage, Charlie (June 27, 2009). "A Bill Signing, With Reservations" (http://thecaucus.blogs.nytimes.com/2009/06/26/a-bill-signing-with-reservations/). *The New York Times*. . Retrieved June 27, 2009.

[62] Woodward, Calvin (April 1, 2009). "PROMISES, PROMISES: Obama tax pledge up in smoke" (http://seattletimes.nwsource.com/html/businesstechnology/2008961613_apobamataxpromise.html). *Seattle Times*. . Retrieved May 24, 2011.

[63] "Obama's new slogan: 'We can't wait'" (http://content.usatoday.com/communities/theoval/post/2011/10/obamas-new-slogan-we-cant-wait/1). *USA Today*. October 24, 2011. .

[64] "'We can't wait': Obama to use executive authority to boost economy as Republicans dither on jobs" (http://www.dailykos.com/story/2011/10/24/1029465/-We-cant-wait:-Obama-to-use-executive-authority-to-boost-economy-as-Republicans-dither-on-jobs). .

[65] "Barack Obama's initial approval rating is highest since JFK" (http://latimesblogs.latimes.com/washington/2009/01/obama-approval.html). *Los Angeles Times*. January 27, 2009. . Retrieved January 28, 2009.

[66] "Obama Approval Ranks Near Top" (http://firstread.msnbc.msn.com/archive/2009/01/26/1760773.aspx). msnbc.com. . Retrieved January 28, 2009. Obama was behind John F. Kennedy at 72%, and above Dwight D. Eisenhower's 68%. He was also behind presidents who came to office in a time of crisis (Harry Truman at 87%, Lyndon B. Johnson at 78%, and Gerald Ford at 71%).

[67] Montopoli, Brian (February 5, 2009). "Obama's Approval Rating: 62 Percent" (http://www.cbsnews.com/blogs/2009/02/05/politics/politicalhotsheet/entry4778267.shtml). *CBS News*. . Retrieved February 5, 2009.

[68] Memmott, Mark (January 30, 2009). "President's approval rating ends week at 66%" (http://content.usatoday.com/communities/theoval/post/2009/01/62157136/1). *USA Today*. . Retrieved February 4, 2009.

[69] "Obama's disapproval rating hits new high" (http://articles.cnn.com/2011-09-13/politics/cnn.poll.obama_1_disapproval-rating-president-barack-obama-handling?_s=PM:POLITICS). CNN. 13 September 2011. . Retrieved 14 May 2012.

[70] "Obama Approval Rating Increases to 67%" (http://www.gallup.com/poll/116224/Obama-Approval-Rating-Increases.aspx). Gallup.com. 2009-02-27. . Retrieved 2010-12-18.

[71] "Obama Approval Index History" (http://www.rasmussenreports.com/public_content/politics/obama_administration/obama_approval_index_history). *Rasmussen Reports*. . Retrieved March 18, 2010.

[72] "Obama's Afghanistan Rating Declines A Year Out, Widespread Anti-Incumbent Sentiment" (http://people-press.org/reports/pdf/561.pdf). November 11, 2009. . Retrieved 2010-02-20.

[73] "Obama Approval Numbers Mirror Stimulus Debate" (http://www.rasmussenreports.com/public_content/most_recent_videos2/business/obama_approval_numbers_mirror_stimulus_debate), *Rasmussen Reports*, February 18, 2009. Retrieved March 18, 2010.

[74] "Obama, Geithner Get Low Grades From Economists" (http://online.wsj.com/article/SB123671107124286261.html). *The Wall Street Journal*. March 11, 2009. Retrieved March 13, 2009.

[75] "Americans Most Confident in Obama on Economy" (http://www.gallup.com/poll/117415/Americans-Confident-Obama-Economy.aspx). *Gallup Poll*. . Retrieved April 17, 2009.

[76] "Obama Rated Highest as Person, Lowest on Deficit, Spending" (http://www.gallup.com/poll/120770/Obama-Rated-Highest-as-Person-Lowest-Deficit-Spending.aspx). *Gallup Poll*. . Retrieved June 15, 2009.

[77] Hechtkopf, Kevin (September 1, 2009). "Poll: Obama Approval Rating Dips Slightly" (http://www.cbsnews.com/blogs/2009/09/01/politics/politicalhotsheet/entry5280405.shtml). *CBS News*. . Retrieved September 9, 2009.

[78] "45% Now Rate Obama's Economic Performance As Poor" (http://www.rasmussenreports.com/public_content/politics/obama_administration/november_2009/45_now_rate_obama_s_economic_performance_as_poor). *Rasmussen Reports*. November 12, 2009. . Retrieved February 20, 2010.

[79] McCormick, John (March 24, 2010), "Obama Widens Approval Above All as Democrats, Republicans Lag" (http://www.bloomberg.com/apps/news?pid=20601070&sid=amj6Ao_aclcE/), *Bloomberg*. Retrieved March 24, 2010.

[80] Blanton, Dana (April 8, 2010). "Fox News Poll: Health Care and the Midterm Elections" (http://www.foxnews.com/politics/2010/04/08/fox-news-poll-health-care-midterm-elections/#). *Fox News*. . Retrieved April 10, 2010.

[81] Blanton, Dana (April 9, 2010). "Fox News Poll: 53% Disapprove of Obama's Handling of the Economy" (http://www.foxnews.com/politics/2010/04/09/fox-news-poll-disapprove-obamas-handling-economy/?test=latestnews). *Fox News*. . Retrieved April 10, 2010.

[82] Gallup Daily: Obama Job Approval (http://www.gallup.com/poll/113980/gallup-daily-obama-job-approval.aspx), *Gallup Poll*, April 10, 2010, Retrieved April 12, 2010 Each result is based on a three-day rolling average

[83] Daily Presidential Tracking Poll (http://www.rasmussenreports.com/public_content/politics/obama_administration/daily_presidential_tracking_poll), *Rasmussen Reports*, April 10, 2010. Retrieved April 12, 2010.

[84] "Election Other - President Obama Job Approval" (http://www.realclearpolitics.com/epolls/other/president_obama_job_approval-1044.html). RealClearPolitics. . Retrieved 2010-12-18.

[85] "Obama Approval Ratings Nationally" (http://www.nationalpolls.com/obama/national.html). *Obama Approval Ratings*. January 3, 2012. . Retrieved January 3, 2012.

[86] http://hdl.loc.gov/loc.uscongress/legislation.111hr847

[87] http://www.gpo.gov/fdsys/pkg/PLAW-111publ353/content-detail.html

[88] http://hdl.loc.gov/loc.uscongress/legislation.111hr2751

[89] Daniel Guarnera (December 19, 2008). "Obama's Cabinet Full ... Except for Drug Czar" (http://www.addictionpro.com/ME2/dirmod.asp?sid=440B847037BC4AFA8B377E381E9C548D&nm=&type=Blog&mod=BlogTopics&mid=67D6564029914AD3B204AD35D8F5F780&tier=7&id=908F70B4E94849F8B3FAA3025FE9221A). . Retrieved January 28, 2009.

[90] "The Cabinet" (http://www.whitehouse.gov/administration/cabinet/). The White House.

[91] "The Obama Cabinet: Confirmations & Nominations" (http://www.npr.org/templates/story/story.php?storyId=99674016). NPR. January 22, 2009. . Retrieved January 28, 2009.

[92] Rushing, J. Taylor (January 19, 2009). "Senate considers unanimous-consent on Obama Cabinet" (http://thehill.com/homenews/news/17822-senate-considers-unanimous-consent-on-obama-cabinet). *The Hill*. . Retrieved January 28, 2009.

[93] "More Obama Cabinet Nominees Confirmed" (http://thecaucus.blogs.nytimes.com/2009/01/22/more-obama-cabinet-nominees-confirmed/). *The New York Times*. January 22, 2009. . Retrieved 2012-01-22.

[94] Knowlton, Brian (January 21, 2009). "On His First Full Day, Obama Tackles Sobering Challenges" (http://www.nytimes.com/2009/01/22/us/politics/22obamacnd.html). *The New York Times* (The New York Times Company). . Retrieved January 28, 2009.

[95] Phillips (January 22, 2009). "More Obama Cabinet Nominees Confirmed" (http://thecaucus.blogs.nytimes.com/2009/01/22/more-obama-cabinet-nominees-confirmed/). *The New York Times* (The New York Times Company). . Retrieved January 28, 2009.

[96] "Geithner confirmed, sworn in at Treasury: Obama cites urgency of acting quickly." (http://www.boston.com/news/nation/washington/articles/2009/01/27/geithner_confirmed_sworn_in_at_treasury/). *The Boston Globe* (The New York Times Company). January 27, 2008. . Retrieved January 28, 2009.

[97] Weisman, Jonathan (January 24, 2009). "Delays in Cabinet Nominations Demonstrate GOP Resolve" (http://online.wsj.com/article/SB123275672929711875.html). *The Wall Street Journal* (Dow Jones & Company, Inc.). . Retrieved January 28, 2009.

[98] Josh Hakala. "Senate confirms Holder as first black AG" (http://www.mlive.com/politics/index.ssf/2009/02/senate_confirms_eric_holder_as.html). Mlive.com. . Retrieved 2012-04-14.

[99] Zeleny, Jeff (February 2, 2009). "Obama Set to Add Republican to Cabinet" (http://www.nytimes.com/2009/02/03/us/politics/03gregg.html). *The New York Times*. . Retrieved February 3, 2009.

[100] Associated Press (6 March 2009). "Tom Daschle Withdraws Nomination For HHS Secretary" (http://www.huffingtonpost.com/2009/02/03/tom-daschle-withdraws-nom_n_163557.html). *The Huffington Post*. . Retrieved 14 May 2012.

[101] "Solis confirmed – CNN Political Ticker" (http://politicalticker.blogs.cnn.com/2009/02/24/solis-confirmed). CNN. 2012-04-10. . Retrieved 2012-04-14.

[102] "Former Dallas Mayor Ron Kirk confirmed as US trade representative" (http://www.chicagotribune.com/news/politics/sns-ap-kirk-trade,0,5324728.story). Chicagotribune.com. . Retrieved 2012-04-14.

[103] Zajac, Andrew (February 4, 2009). "Commerce Nominee's Own Finances Have Suffered". Chicago Tribune.

[104] Rachel Weiner (15 March 2009). "Judd Gregg Withdraws Commerce Secretary Nomination (VIDEO)" (http://www.huffingtonpost.com/2009/02/12/judd-gregg-withdraws-comm_n_166502.html). *The Huffington Post*. . Retrieved 14 May 2012.

[105] Christi Parsons (9 March 2011). "Gary Locke, Commerce secretary, nominated as U.S. ambassador to China" (http://articles.latimes.com/2011/mar/09/news/la-pn-locke-china-ambassador-20110310). *Los Angeles Times*. . Retrieved 14 May 202.

[106] Wheaton, Sarah (March 24, 2009). "Locke Confirmed as Commerce Secretary" (http://thecaucus.blogs.nytimes.com/2009/03/24/locke-confirmed-as-commerce-secretary/). *The New York Times*. . Retrieved March 30, 2009.

[107] "President Obama Will Nominate Governor Kathleen Sebelius Secretary of HHS, Announces Release of $155 Million of ARRA Funds for Health Clinics Across America" (http://www.whitehouse.gov/the_press_office/President-Obama-nominates-Governor-Kathleen-Sebelius-Secretary-of-HHS-Announces-Re/). White House. 2 March 2009. . Retrieved 14 May 2012.

[108] Associated Press (13 May 2009). "Obama To Tap Sebelius for HHS Secretary" (http://www.cbsnews.com/2100-250_162-4835996.html). CBS News. .

[109] Obama's promise of ethics reform faces early test (http://www.iht.com/articles/2009/02/03/america/03lobby.php), International Herald Tribune, February 3, 2009

[110] Tax issues prompt Obama nominee to withdraw (http://articles.cnn.com/2009-02-03/politics/performance.nominee.withdraws_1_nancy-killefer-chief-performance-officer-unemployment-tax?_s=PM:POLITICS) (February 3, 2009), CNN.

[111] Tax problems delay vote on Hilda Solis (http://www.upi.com/Top_News/2009/02/05/Tax_problems_delay_vote_on_Hilda_Solis/UPI-61081233871528/), UPI Online, February 5, 2009

[112] "Trade nominee Ron Kirk agrees to pay back taxes" (http://www.bostonherald.com/business/general/view.bg?articleid=1155758&srvc=next_article). Bostonherald.com. 2009-03-02. . Retrieved 2012-04-14.

[113] Another Obama Nominee Has Tax Issues (http://www.cbsnews.com/blogs/2009/03/31/politics/politicalhotsheet/entry4908247.shtml), CBS News, March 31, 2009

[114] Michael A. Fletcher, Obama Criticized as Too Cautious, Slow on Judicial Posts (http://www.washingtonpost.com/wp-dyn/content/article/2009/10/15/AR2009101504083.html), *The Washington Post* (October 16, 2009).

[115] "Leahy Blasts Republicans For Stalled Nominees" (http://www.mainjustice.com/2009/12/15/leahy-blasts-gop-for-stalled-nominees/). Washington Time. . Retrieved 2009-12-16.

[116] Millhiser, Ian (July 30, 2010). "Falling Off a Cliff: Judicial Confirmation Rates Have Nosedived in the Obama Presidency" (http://www.americanprogress.org/issues/2010/07/judicial_confirmations.html). Center for American Progress. . Retrieved 2010-08-12.

[117] Henry, Ed (January 19, 2009). "Obama's top priority: the economy" (http://money.cnn.com/2009/01/19/news/economy/obama_first_week/). CNN. . Retrieved February 9, 2009.

[118] Goldman, Russell; Jake Tapper (January 5, 2009). "Obama Pushes Economic Plan, Saying It Can't Wait" (http://abcnews.go.com/Business/Business/story?id=6568008&page=1). ABC News. . Retrieved February 9, 2009.

[119] Bohan, Caren; Lisa Lambert (February 20, 2009). "Obama vows to spend stimulus money wisely" (http://uk.reuters.com/article/marketsNewsUS/idUKN2044566820090220). Reuters. . Retrieved February 24, 2009.

[120] Halloran, Liz (January 27, 2009). "Obama Woos Hill Republicans On Stimulus Plan" (http://www.npr.org/templates/story/story.php?storyId=99927275). NPR. . Retrieved February 9, 2009.

[121] "Committee for a Responsible Federal Budget, Stimulus Watch" (http://stimulus.org/). .

[122] Condon, Stephanie (April 15, 2010). "What's Obama Doing to Your Taxes?" (http://www.cbsnews.com/8301-503544_162-20002548-503544.html). CBS News. . Retrieved 2010-11-13.

[123] "Budget of the United States Government, Fiscal Year 2011" (http://www.whitehouse.gov/sites/default/files/omb/budget/fy2011/assets/budget.pdf) (PDF). . Retrieved 2012-04-14.

[124] "Labor Force Statistics from the Current Population Survey" (http://data.bls.gov/PDQ/servlet/SurveyOutputServlet?request_action=wh&graph_name=LN_cpsbref3). *United States Department of Labor Bureau of Labor Statistics*. August 1, 2009. .

[125] Stephanopoulos, George (July 5, 2009). "Biden: We 'Misread the Economy'" (http://blogs.abcnews.com/george/2009/07/biden-we-misread-the-economy-.html). *George's Bottom Line* (ABC News). . Retrieved April 20, 2010.

[126] Stephanopoulos, George (July 5, 2009). "'This Week' Transcript: Exclusive: Vice President Joe Biden" (http://abcnews.go.com/ThisWeek/Politics/story?id=8002421&page=3). *This Week* (ABC News): 3. . Retrieved April 20, 2010.

[127] MacGillis, Alec (January 13, 2010). "Economic stimulus has created or saved nearly 2 million jobs, White House says" (http://www.washingtonpost.com/wp-dyn/content/article/2010/01/12/AR2010011203737.html). *The Washington Post*. . Retrieved May 22, 2010.

[128] "Tracking the Money" (http://www.recovery.gov/Pages/home.aspx). Recovery.gov. . Retrieved 2010-12-18.

[129] Theodossiou, Eleni; Hipple, Steven F. (2011). "Unemployment Remains High in 2010" (http://www.bls.gov/opub/mlr/2011/03/art1full.pdf). *Monthly Labor Review* (Bureau of Labor Statistics) **134** (3): 3–22. . Retrieved April 7, 2011.

[130] Eddlemon, John P. (2011). "Payroll Employment Turns the Corner in 2010" (http://www.bls.gov/opub/mlr/2011/03/art2full.pdf). *Monthly Labor Review* (Bureau of Labor Statistics) **134** (3): 23–32. . Retrieved April 7, 2011.

[131] "Percent Change in Real Gross Domestic Product (Quarterly)" (http://bea.gov/national/nipaweb/TableView.asp?SelectedTable=1&ViewSeries=NO&Java=no&Request3Place=N&3Place=N&FromView=YES&Freq=Qtr&FirstYear=2009&LastYear=2010&3Place=N&Update=Update&JavaBox=no#Mid). *National Income and Product Accounts Table*. Bureau of Economic Analysis. . Retrieved April 7, 2011.

[132] "Percent Change in Real Gross Domestic Product (Annual)" (http://bea.gov/national/nipaweb/TableView.asp?SelectedTable=1&ViewSeries=NO&Java=no&Request3Place=N&3Place=N&FromView=YES&Freq=Year&FirstYear=2009&LastYear=2010&3Place=N&Update=Update&JavaBox=no#Mid). *National Income and Product Accounts Table*. Bureau of Economic Analysis. . Retrieved April 7, 2011.

[133] Thoma, Mark (December 1, 2010). "Senate GOP Pledges to Block All Bills Until Tax Cuts are Extended for All" (http://wallstreetpit.com/52023-senate-gop-pledges-to-block-all-bills-until-tax-cuts-are-extended-for-all). Wall Street Pit. . Retrieved 2010-12-20.

[134] "Senate Republicans block 9/11 health bill" (http://www.reuters.com/article/2010/12/10/us-usa-congress-firefighters-idUSTRE6B903120101210). Reuters. December 9, 2010. . Retrieved 2012-5-14.

[135] Stirewalt, Chris (December 1, 2010). "Today's Power Play: Republicans and Democrats Play Chicken With Lame Duck" (http://politics.blogs.foxnews.com/2010/12/01/todays-power-play-republicans-and-democrats-play-chicken-lame-duck). Fox News Channel. . Retrieved 2010-12-20.

[136] Herszenhorn, David M.; Stolberg, Sheryl Gay (December 7, 2010). "Democrats Skeptical of Obama on New Tax Plan" (http://www.nytimes.com/2010/12/08/us/politics/08cong.html). *The New York Times*. .

[137] Nick Wing, Bernie Sanders Filibuster: Senator Stalls Tax Cut Deal (http://www.huffingtonpost.com/2010/12/10/bernie-sanders-filibuster_n_795087.html), December 10, 2010, The Huffington Post.

[138] "Obama signs tax deal into law" (http://www.cnn.com/2010/POLITICS/12/17/tax.deal/index.html). CNN. December 17, 2010. . Retrieved December 17, 2010.

[139] Beutler, Brian (December 16, 2010). "House Passes Tax Cut Plan, Obama To Sign" (http://tpmdc.talkingpointsmemo.com/2010/12/house-passes-tax-cut-plan-obama-to-sign.php). Talking Points Memo. . Retrieved December 20, 2010.

[140] Post Store (December 17, 2010). "Congress votes to extend Bush-era tax cuts until '12" (http://www.washingtonpost.com/wp-dyn/content/article/2010/12/16/AR2010121606672.html). *The Washington Post*. . Retrieved 2010-12-17.

[141] Kosterlitz J. (2009). Obama Aiming To Lock Turnstile For Lobbyists (http://www.nationaljournal.com/njmagazine/sl_20090321_5174.php). *National Journal*.

[142] Obama White House Discloses Two More Lobbyist Waivers Granted (http://blogs.abcnews.com/politicalpunch/2009/03/obama-white-hou.html), ABC News, March 10, 2009

[143] "Ethics Update" (http://www.whitehouse.gov/blog/09/03/10/Ethics-Update/), The White House

[144] Jack Tapper (September 5, 2009). "White House Discloses 10 More Ethics Waivers for Administration Officials" (http://blogs.abcnews.com/politicalpunch/2009/09/white-house-discloses-10-more-ethics-waivers-for-administration-officials.html). .

[145] Schouten, Fredreka (January 27, 2009). "Geithner names ex-lobbyist as Treasury chief of staff" (http://www.usatoday.com/news/washington/2009-01-27-lobbyist_N.htm). *USA Today*. . Retrieved February 14, 2009.

[146] "Obama HHS deputy has recent lobbyist past" (http://www.politico.com/news/stories/0109/17376.html). *Politico*. January 13, 2009. . Retrieved February 14, 2009.

[147] The Nominee Who Lobbied Herself (http://www.weeklystandard.com/Content/Public/Articles/000/000/016/101mswgc.asp), The Weekly Standard, February 3, 2009

[148] Macon Phillips (January 20, 2009). "Change has come to WhiteHouse.gov" (http://www.whitehouse.gov/blog/change_has_come_to_whitehouse-gov/). The White House. . Retrieved January 26, 2009.

[149] Obama breaks five-day pledge (http://www.politico.com/news/stories/0209/18441.html), *Politico*, February 5, 2009

[150] Barack Obama Campaign Promise No. 234: Allow five days of public comment before signing bills (http://politifact.com/truth-o-meter/promises/promise/234/allow-five-days-of-public-comment-before-signing-b/), Politifact, February 4, 2009

[151] We are better than these last eight years (http://www.msnbc.msn.com/id/26446638/wid/7468326/page/5/6.b10984/6.b10984/6.a10984/2??cm=WaterCooler-SC) Prepared remarks of Barack Obama to the Democratic National Convention

[152] "Executive Order 13489 -- Presidential Records" (http://www.whitehouse.gov/the_press_office/ExecutiveOrderPresidentialRecords/). The White House. . Retrieved January 22, 2009.

[153] "New Obama Orders on Transparency, FOIA Requests" (http://voices.washingtonpost.com/federal-eye/2009/01/_in_a_move_that.html). *The Washington Post*. . Retrieved January 22, 2009.

[154] "RELEASED: The Bush Administration's Secret Legal Memos" (http://www.aclu.org/safefree/general/olc_memos.html). American Civil Liberties Union. . Retrieved April 19, 2009.

[155] Vick, Karl (April 25, 2009). "Amid Outcry on Memo, Signer's Private Regret" (http://www.washingtonpost.com/wp-dyn/content/article/2009/04/24/AR2009042403888.html). *The Washington Post*. . Retrieved April 24, 2009.

[156] Hess, Pamela (April 19, 2009). "Four CIA chiefs said 'don't reveal torture memos'" (http://www.independent.co.uk/news/world/americas/four-cia-chiefs-said-dont-reveal-torture-memos-1671068.html). *The Independent* (London). . Retrieved April 19, 2009.

[157] "The Torturers' Manifesto" (http://www.nytimes.com/2009/04/19/opinion/19sun1.html). *The New York Times*. April 18, 2009. . Retrieved April 19, 2009.

[158] Smith, R. Jeffrey (April 19, 2009). "Justice Dept. Memos' Careful Legalese Obscured Harsh Reality" (http://www.washingtonpost.com/wp-dyn/content/article/2009/04/18/AR2009041802166.html). *The Washington Post*. . Retrieved April 19, 2009.

[159] MacAskill, Ewen (May 21, 2009). "Obama and Cheney clash over Guantanamo torture" (http://www.guardian.co.uk/world/2009/may/21/obama-administration-dickcheney). *The Guardian* (London). . Retrieved June 10, 2009.

[160] http://www.recovery.gov/

[161] "H.R.1 - American Recovery and Reinvestment Act of 2009" (http://www.opencongress.org/bill/111-h1/text). OpenCongress.org. . Information cited is found under Subtitle B on pages 15-18.

[162] "Obama blocks list of visitors to White House" (http://news.mobile.msn.com/en-us/articles.aspx?afid=1&aid=31373407). MSNBC. 16 June 2009. . Retrieved 14 May 2012.

[163] "White House Voluntary Disclosure Policy Visitor Access Records" (http://www.whitehouse.gov/VoluntaryDisclosure/). The White House. . Retrieved September 4, 2009.

[164] "White House Releases October Visitor Records" (http://whitehouse.blogs.foxnews.com/2010/01/29/white-house-releases-october-visitor-records/). Fox News. January 29, 2010. . Retrieved April 19, 2010.

[165] "White House Posts More than 75,000 Visitors Records" (http://blogs.abcnews.com/politicalpunch/2010/01/white-house-posts-more-than-75000-visitors-records----including-farrakhan-and-james-taylor-no-not-the-ones-youre-thinking.html). ABC News. January 29, 2010. . Retrieved April 19, 2010.

[166] "More Than 250,000 White House Visitor Records Now Online" (http://www.whitehouse.gov/blog/2010/03/26/more-than-250000-white-house-visitor-records-now-online). The White House. March 26, 2010.

[167] "Barack Obama Campaign Promise No. 517" (http://www.politifact.com/truth-o-meter/promises/promise/517/health-care-reform-public-sessions-C-SPAN/). Politifact. July 10, 2009. . Retrieved April 21, 2010.

[168] Markay, Lachlan (January 8, 2010). "Gibbs Refuses to Address Question About Obama's Broken C-SPAN Promises" (http://online.wsj.com/article/SB10001424052748703481004574646560762972536.html). *The Wall Street Journal*. . Retrieved April 21, 2010.

[169] "Obama acknowledges broken C-SPAN promise" (http://www.washingtontimes.com/news/2010/feb/02/obama-acknowledges-broken-c-span-promise/). *The Washington Times*. February 2, 2010. . Retrieved April 21, 2010.

[170] Scott Shane (June 11, 2010). "Obama Takes a Hard Line Against Leaks to Press" (http://www.nytimes.com/2010/06/12/us/politics/12leak.html). *The New York Times*. .

[171] Mark Hosenball (April 16, 2010). "Exclusive: House Republican Staffer Introduced Alleged NSA Leaker to Reporter" (http://www.thedailybeast.com/newsweek/blogs/declassified/2010/04/16/exclusive-house-republican-staffer-introduced-alleged-nsa-leaker-to-reporter.html). *Newsweek*. . Retrieved February 10, 2012.

[172] "Setback in case against accused NSA leaker" (http://voices.washingtonpost.com/checkpoint-washington/2010/11/setback_in_case_against_accuse.html), Ellen Nakashima, *The Washington Post*, November 29, 2010, Retrieved March 10, 2011

[173] Indictment Continues Obama Administration's War on Leaks (http://www.washingtonian.com/blogarticles/18114.html) ,Shane Harris, washingtonian, 01/25/2011, Retrieved March 9, 2011

[174] "Classified-Info Crackdown" (http://www.newsweek.com/blogs/declassified/2010/06/11/classified-info-crackdown.html), Michael Isikoff, Newsweek, June 11, 2010, Retrieved March 11, 2011

[175] "U.S. Analyst Is Indicted in Leak Case" (http://www.nytimes.com/2010/08/28/world/americas/28leak.html), August 27, 2010, Scott Shane, *The New York Times*, Retrieved March 11, 2011

[176] Savage, Charlie (28 April 2010). "U.S. Subpoenas Times Reporter Over Book on C.I.A." (http://www.nytimes.com/2010/04/29/us/29justice.html). *The New York Times*. . Retrieved 12 March 2012.

[177] Macon Phillips (January 20, 2009). "President Barack Obama's Inaugural Address" (http://www.whitehouse.gov/blog/inaugural-address/). The White House. . Retrieved January 26, 2009.

[178] Baldwin, Tom (January 22, 2009). "President Obama's first call 'was to President Abbas'" (http://www.timesonline.co.uk/tol/news/world/us_and_americas/us_elections/article5563280.ece). *The Times* (London). . Retrieved January 23, 2009.

[179] "President Obama Delivers Remarks to State Department Employees" (http://www.washingtonpost.com/wp-dyn/content/article/2009/01/22/AR2009012202550.html). *The Washington Post*. January 22, 2009. . Retrieved January 23, 2009.

[180] "Obama urges Israel to open Gaza borders" (http://www.ft.com/cms/s/0/7cf745dc-e8ce-11dd-a4d0-0000779fd2ac.html). . Retrieved January 22, 2009.

[181] Hodge, Amanda (February 19, 2009). "Obama launches Afghanistan surge" (http://www.theaustralian.news.com.au/story/0,25197,25074581-2703,00.html). The Australian. .

[182] "Gates: More Troops For Afghanistan" (http://www.news10.net/news/national/story.aspx?storyid=53827). news10.net. January 27, 2009. .

[183] "U.S. general urges troop surge in Afghanistan" (http://www.iht.com/articles/2008/10/01/mideast/military.php). *International Herald Tribune*. October 1, 2008. .

[184] "Obama sets date to end Iraq combat mission" (http://www.msnbc.msn.com/id/29371588/ns/world_news-mideast_n_africa/t/obama-sets-date-end-iraq-combat-mission/). MSNBC. February 27, 2009. .

[185] Ewen MacAskill (27 February 2009). "Six years after Iraq invasion, Obama sets out his exit plan" (http://www.guardian.co.uk/world/2009/feb/27/obama-iraq-war-end-august-2010). *The Guardian*. . Retrieved 14 May 2012.

[186] Mathiason, Nick (March 1, 2009). "So long, farewell, auf wiedersehen: is it game over for Swiss banks" (http://www.guardian.co.uk/business/2009/mar/01/swiss-banks-anonymity-ubs). The Guardian. London. .

[187] Smith, R. Jeffrey; Candace Rondeaux; Joby Warrick (January 24, 2009). "2 U.S. Airstrikes Offer a Concrete Sign of Obama's Pakistan Policy" (http://www.washingtonpost.com/wp-dyn/content/article/2009/01/23/AR2009012304189.html). *The Washington Post*. .

[188] Heilprin, John (January 26, 2009). "Obama Administration To Engage In "Direct Diplomacy" With Iran" (http://www.huffingtonpost.com/2009/01/26/direct-diplomacy-with-ira_n_160952.html). *Huffington Post*. Associated Press. .

[189] "Ambassador Bosworth To Lead Efforts on U.S. Engagement With North Korea" (http://www.state.gov/secretary/rm/2009a/02/119861.htm). U.S. Department of State. February 26, 2009. .

[190] "Statement On Bilateral Meeting With President Hu Of China" (http://www.whitehouse.gov/the_press_office/Statement-On-Bilateral-Meeting-With-President-Hu-Of-China/). The White House. 2009-04-01. . Retrieved 2010-02-20.

[191] Barack Obama uses Bush funding tactics to finance wars in Iraq and Afghanistan (http://www.telegraph.co.uk/news/worldnews/northamerica/usa/barackobama/5133899/Barack-Obama-uses-Bush-funding-tactics-to-finance-wars-in-Iraq-and-Afghanistan.html), The Telegraph, April 10, 2009

[192] Pentagon plans 20,000 new jobs to manage arms-buys (http://www.reuters.com/article/domesticNews/idUSTRE54545N20090506), Reuters, May 6, 2009

[193] Jesse Lee (June 3, 2009). "The President in the Middle East" (http://www.whitehouse.gov/blog/The-President-in-the-Middle-East/). White House. . Retrieved June 4, 2009.

[194] "Obama in Egypt reaches out to Muslim world - CNN.com" (http://www.cnn.com/2009/POLITICS/06/04/egypt.obama.speech/index.html). *CNN*. June 4, 2009. . Retrieved May 22, 2010.

[195] Zeleny, Jeff; Cowell, Alan (June 5, 2009). "Addressing Muslims, Obama Pushes Mideast Peace" (http://www.nytimes.com/2009/06/05/world/middleeast/05prexy.html). *The New York Times*. . Retrieved May 22, 2010.

[196] Associated Press (4 June 2009). "Obama speech widely hailed but foes still sceptical" (http://www.google.com/hostednews/afp/article/ALeqM5hOFrIyS2OYEJuBwjdcFTeFfJQ7ng). Google News. . Retrieved 14 May 2012.

[197] "Middle East | Reaction: Obama's Cairo speech" (http://news.bbc.co.uk/2/hi/middle_east/8083171.stm). BBC News. June 4, 2009. . Retrieved June 5, 2009.

[198] Israel says shares Obama's hopes for peace (http://www.ynetnews.com/articles/0,7340,L-3726482,00.html), Yedioth Ahronoth, Attila Somfalvi , June 4, 2009

[199] Spencer, Richard (May 6, 2009). "Barack Obama's speech to Muslim world welcomed by the press" (http://www.telegraph.co.uk/news/worldnews/northamerica/usa/barackobama/5450146/Barack-Obamas-speech-to-Muslim-world-welcomed-by-the-press.html). *The Telegraph* (London). . Retrieved June 5, 2009.

[200] "Pakistan backs Obama's Middle East approach" (http://thenews.jang.com.pk/daily_detail.asp?id=181395). Thenews.jang.com.pk. November 26, 2008. . Retrieved June 6, 2009.

[201] "Clinton rebukes Israel over homes" (http://news.bbc.co.uk/2/hi/middle_east/8565455.stm). *BBC News*. March 12, 2010. . Retrieved May 22, 2010.

[202] "Obama, Russian President Sign Arms Treaty" (http://www.cnn.com/2010/POLITICS/04/08/obama.russia.treaty/index.html). CNN. April 8, 2010. . Retrieved April 8, 2010.

[203] "Gaddafi 'not targeted' by allied strikes" (http://www.bbc.co.uk/news/world-africa-12813757). BBC News, March 21, 2011.

[204] Finn, Peter (January 21, 2009). "Obama Seeks Halt to Legal Proceedings at Guantanamo" (http://www.washingtonpost.com/wp-dyn/content/article/2009/01/20/AR2009012004743_pf.html). *The Washington Post*. Associated Press. . Retrieved January 21, 2009.

[205] "Obama signs executive order to close Guantanamo Bay" (http://edition.cnn.com/2009/POLITICS/01/22/guantanamo.order/index.html). CNN. January 22, 2009. . Retrieved January 22, 2009.

[206] Obama, Barack (January 22, 2009). "Executive Order 13491 – Ensuring Lawful Interrogations" (http://www.whitehouse.gov/the_press_office/EnsuringLawfulInterrogations/). The White House. . Retrieved January 26, 2009.

[207] Guantanamo worse since Obama election: ex-detainee (http://www.google.com/hostednews/afp/article/ALeqM5gyQRvqsH3BufMb21Gv8banWRn54A), AFP, March 7, 2009

[208] Glaberson, William (March 13, 2009). "U.S. Won't Label Terror Suspects as 'Combatants'" (http://www.nytimes.com/2009/03/14/us/politics/14gitmo.html). *The New York Times*. . Retrieved March 15, 2009.

[209] Loven, Jennifer (September 27, 2009). "AP sources: Gitmo closing goal of January may slip" (http://www.guardian.co.uk/world/feedarticle/8726455). *The Guardian*. Associated Press (London). . Retrieved February 9, 2012.

[210] "Obama Admits Closing Date Will Slip" (http://news.bbc.co.uk/2/hi/americas/8366376.stm). BBC News. November 18, 2009. . Retrieved November 22, 2009.

[211] "Public Hearing for Moving Guantanamo Detainees to Illinois" (http://www.chicagopublicradio.org/Content.aspx?audioID=39050). Chicago Public Radio. December 23, 2009. . Retrieved January 23, 2010.

[212] Mazzetti, Mark; Helene Cooper, Peter Baker (3 May 2011). "Clues Gradually Led to the Location of Osama bin Laden" (http://www.nytimes.com/2011/05/03/world/asia/03intel.html?pagewanted=all). *The New York Times*. . Retrieved 4 May 2011.

[213] "Osama bin Laden buried at sea after being killed by U.S. forces in Pakistan" (http://www.washingtonpost.com/politics/osama-bin-laden-is-killed-by-us-forces-in-pakistan/2011/05/01/AFXMZyVF_story.html). *The Washington Post*. May 2, 2011

[214] "Official offers details of bin Laden raid" - newsday.com (http://www.newsday.com/news/breaking/official-offers-details-of-bin-laden-raid-1.2853079) Retrieved May 2, 2011

[215] Schabner, Dean; Karen Travers (1 May 2011). "Osama bin Laden Killed by U.S. Forces in Pakistan" (http://abcnews.go.com/Blotter/osama-bin-laden-killed/story?id=13505703). ABC News. . Retrieved 3 May 2011.

[216] Baker, Peter; Helene Cooper; Mark Mazzetti (2 May 2011). "Bin Laden Is Dead, Obama Says" (http://www.nytimes.com/2011/05/02/world/asia/osama-bin-laden-is-killed.html). *The New York Times*. . Retrieved 3 May 2011.

[217] Walsh, Declan; Richard Adams and Ewen MacAskill (May 2, 2011). "Osama bin Laden is dead, Obama announces" (http://www.guardian.co.uk/world/2011/may/02/osama-bin-laden-dead-obama). *The Guardian* (London). . Retrieved May 3, 2011.

[218] Dorning, Mike (May 2, 2011). "Death of Bin Laden May Strengthen Obama's Hand in Domestic, Foreign Policy" (http://www.bloomberg.com/news/2011-05-02/death-of-bin-laden-may-strengthen-obama-s-hand-in-domestic-foreign-policy.html). Bloomberg News. . Retrieved 4 May 2011.

[219] "World Reaction To Osama Bin Laden's Death" (http://www.npr.org/2011/05/02/135919728/world-reaction-to-osama-bin-ladens-death). NPR. May 2, 2011. . Retrieved 4 May 2011.

[220] "Inaugural Address" (http://abcnews.go.com/Politics/Inauguration/story?id=6689022&page=1). Abcnews.go.com. 2009-01-20. . Retrieved 2012-04-14.

[221] Shane, Scott (April 6, 2010). "U.S. Approves Targeted Killing of American Cleric" (http://www.nytimes.com/2010/04/07/world/middleeast/07yemen.html). *The New York Times*. . Retrieved May 22, 2010.

[222] Becker, Jo; Shane, Scott (May 29, 2012). "Secret 'Kill List' Proves a Test of Obama's Principles and Will" (http://www.nytimes.com/2012/05/29/world/obamas-leadership-in-war-on-al-qaeda.html). *The New York Times*: pp. 7–8. . Retrieved June 14, 2012.

[223] "NRA Targets Obama" (http://www.factcheck.org/elections-2008/nra_targets_obama.html), *FactCheck*, September 22, 2008 (updated September 29, 2008). Retrieved March 15, 2010.

[224] "Barack Obama on Gun Control" (http://www.ontheissues.org/domestic/Barack_Obama_Gun_Control.htm), *On the Issues*. Retrieved March 15, 2010.

[225] Ryan, Jason (February 25, 2009). "Obama to Seek New Assault Weapons Ban" (http://abcnews.go.com/print?id=6960824). ABC News. . Retrieved February 26, 2009.

[226] Isikoff, Michael; Smalley, Suzanne (April 11, 2009). "Back Burner for an Assault-Weapon Ban" (http://www.newsweek.com/id/193589). *Newsweek*. . Retrieved March 15, 2010.

[227] "Senate Votes to Allow Guns on Amtrak" (http://www.cbsnews.com/stories/2009/09/16/national/main5316093.shtml), CBS News, September 16, 2009. Retrieved March 15, 2010.

[228] Urbina, Ian (February 23, 2010), "Fearing Obama Agenda, States Push to Loosen Gun Laws" (http://www.nytimes.com/2010/02/24/us/24guns.html), *The New York Times*. Retrieved March 15, 2010.

[229] Dinan, Stephen (February 22, 2010), "Parks open to holders of concealed guns" (http://www.washingtontimes.com/news/2010/feb/22/national-parks-will-open-gates-to-holders-of-conce/), *The Washington Times*. Retrieved March 15, 2010.

[230] Hathaway, Melissa (April 22, 2009). "Remarks by Melissa E. Hathaway Acting Senior Director for Cyberspace for the National Security and Homeland Security Councils" (http://voices.washingtonpost.com/securityfix/Melissa Hathaway Speech at RSA.pdf). *The Washington Post*. . Retrieved April 22, 2009.

[231] "President Obama Directs the National Security and Homeland Security Advisors to Conduct Immediate Cyber Security Review" (http://www.whitehouse.gov/the_press_office/advisorstoconductimmediatecybersecurityreview/) (Press release). The White House. February 9, 2009. . Retrieved April 17, 2009.

[232] Molina, Brett (April 15, 2009). "Obama to receive cybersecurity review this week" (http://content.usatoday.com/communities/technologylive/post/2009/04/65504587/1#.T7GD1dmkBpQ). *USA Today*. . Retrieved May 14, 2012.

[233] Lichtblau, Eric and Risen, James (April 15, 2009). "N.S.A.'s Intercepts Exceed Limits Set by Congress" (http://www.nytimes.com/2009/04/16/us/16nsa.html). *The New York Times*. . Retrieved April 15, 2009.

[234] Ackerman, Spencer (April 16, 2009). "NSA Revelations Spark Push to Restore FISA" (http://washingtonindependent.com/39153/nsa-revelations-spark-movement-to-restore-fisa). *The Washington Independent* (Center for Independent Media). . Retrieved April 19, 2009.

[235] Mufson, Steven; Juliet Eilperin (January 27, 2009). "Obama Issues Orders Toward More Fuel-Efficient Cars" (http://www.washingtonpost.com/wp-dyn/content/article/2009/01/26/AR2009012602028.html). *The Washington Post*. . Retrieved January 27, 2009.

[236] "American Recovery and Reinvestment Act of 2009" (http://web.archive.org/web/20110106095705/http://waxman.house.gov/UploadedFiles/ARRA.pdf). 20 April 2009. . Retrieved 14 May 2012.

[237] "Bush-era offshore drilling plan is set aside" (http://www.msnbc.msn.com/id/29119940). MSNBC. . Retrieved 2012-04-14.

[238] Broder, John M. (May 19, 2009). "Obama to Toughen Rules on Emissions and Mileage" (http://www.nytimes.com/2009/05/19/business/19emissions.html). *The New York Times*. . Retrieved May 22, 2010.

[239] Broder, John M. (March 30, 2010). "Obama to Open Offshore Areas to Oil Drilling for First Time" (http://www.nytimes.com/2010/03/31/science/earth/31energy.html). *The New York Times*. . Retrieved April 7, 2010.

[240] Baker, Peter (2010-05-27). "Obama Extends Moratorium; Agency Chief Resigns" (http://www.nytimes.com/2010/05/28/us/28drill.html). *New York Times*. . Retrieved 2010-06-08.

[241] "Obama, in Gulf, pledges to push on stopping leak" (http://www.usatoday.com/news/nation/2010-05-27-oil-spill-news_N.htm). *USA Today*. 2010-05-27. . Retrieved 2010-05-27.

[242] Hechtkopf, Kevin (2010-04-30). "Obama Promises Review of Gulf Oil Spill - Political Hotsheet" (http://www.cbsnews.com/8301-503544_162-20003862-503544.html). CBS News. . Retrieved 2010-06-06.

[243] James Oliphant; Peter Nicholas (2010-06-01). "Gulf oil spill: Obama administration launches criminal probe of rig explosion" (http://latimesblogs.latimes.com/greenspace/2010/06/gulf-oil-spill-obama-administration-launches-criminal-probe-of-rig-explosion-.html). *Los Angeles Times*. . Retrieved 2010-06-06.

[244] Etter, Lauren (June 1, 2010). "U.S. Opens Criminal Probe on Spill" (http://online.wsj.com/article/SB10001424052748704875604575280363277341150.html). *The Wall Street Journal*. . Retrieved 2010-06-06.

[245] Etter, Lauren (2010-06-08). "Obama, on 'Today,' wonders 'whose ass to kick' following BP oil spill" (http://www.mlive.com/politics/index.ssf/2010/06/obama_on_today_wonders_whose_a.html). Mlive.com. . Retrieved 2010-06-08.

[246] "Gulf oil spill: Obama's big political test" (http://www.csmonitor.com/USA/Politics/2010/0529/Gulf-oil-spill-Obama-s-big-political-test). *The Christian Science Monitor*. 2010-05-29. . Retrieved 2010-06-06.

[247] Amos, Jonathan (February 1, 2010). "Obama cancels Moon return project" (http://news.bbc.co.uk/1/hi/sci/tech/8489097.stm). *BBC News*. . Retrieved March 7, 2010.

[248] "Terminations, Reductions, and Savings" (http://www.whitehouse.gov/sites/default/files/omb/budget/fy2011/assets/trs.pdf). . Retrieved March 7, 2010.

[249] Achenbach, Joel (February 1, 2010). "NASA budget for 2011 eliminates funds for manned lunar missions" (http://www.washingtonpost.com/wp-dyn/content/article/2010/01/31/AR2010013101058.html). *The Washington Post*. . Retrieved February 1, 2010.

[250] Obama scraps funding for Nasa's American Moon base mission (http://www.bbc.co.uk/worldservice/news/2010/02/100201_obama_moon_wup_sl.shtml). BBC

[251] Art Chimes. "Expert Group Says NASA Budget Too Small for Big Space Plans" (http://www.voanews.com/english/news/a-13-2009-08-26-voa54-68820852.html). VOA News. . Retrieved September 1, 2009.

[252] Clara Moskowitz (2010-06-24). "Misconceptions swirl around Obama space plan" (http://www.msnbc.msn.com/id/37907917/ns/technology_and_science-space/t/misconceptions-swirl-around-obama-space-plan/). msnbc.com. . Retrieved 2010-07-07.

[253] "Obama to unveil 'ambitious plan' for NASA" (http://www.msnbc.msn.com/id/35751461/ns/technology_and_science-space/). Associated Press. msnbc.com. 2010-03-07. . Retrieved 2010-05-18.

[254] "Review of U.S. Plans Committee" (http://www.nasa.gov/pdf/396093main_HSF_Cmte_FinalReport.pdf). Human Space Flights Committee. . Retrieved 2010-07-07.

[255] "Federal Budget Fiscal Year 2011 - Fact Sheet on Innovation" (http://www.whitehouse.gov/omb/factsheet_key_innovation/). Office of Management and Budget. 2010. . Retrieved 2010-06-15.

[256] "Ex-astronaut Bolden to lead Nasa" (http://news.bbc.co.uk/2/hi/science/nature/8154525.stm). BBC. 2009-07-16. .

[257] "Obama overturns Bush policy on stem cells" (http://www.cnn.com/2009/POLITICS/03/09/obama.stem.cells/index.html). Cnn.com. 2009-03-09. . Retrieved 2012-04-14.

[258] " Obama Ends Global Family Planning Restrictions (http://www.npr.org/templates/story/story.php?storyId=99831044&ft=1&f=1001)". NPR: National Public Radio. January 23, 2009.

[259] Bailey, Holly (June 17, 2009). "Obama: LGBT Benefits Memo "Just a Start"" (http://blog.newsweek.com/blogs/thegaggle/archive/2009/06/17/obama-lgbt-benefits-memo-just-a-start.aspx). *Newsweek*. . Retrieved June 17, 2009.

[260] Ira Kantor (June 24, 2009). "Gay rights advocates protest Dem fundraiser" (http://www.bostonherald.com/news/regional/view/20090624gay_rights_advocates_protest_dem_fundraiser/). *Boston Herald*. . Retrieved 2010-12-18.

[261] "Donors say Obama slow on pledge to gays" (http://www.msnbc.msn.com/id/31550347/ns/politics-white_house/). msnbc.com. 2009-06-25. . Retrieved 2010-12-18.

[262] Office of Public Affairs (October 19, 2009). "Attorney General announces formal medical marijuana guidelines" (http://www.justice.gov/opa/pr/2009/October/09-ag-1119.html). U.S. Department of Justice. . Retrieved July 19, 2010.

Ogden, David W. (October 19, 2009). "Memorandum for selected United States Attorneys: Investigations and prosecutions in states authorizing the medical use of marijuana" (http://www.justice.gov/opa/documents/medical-marijuana.pdf). U.S. Department of Justice. . Retrieved July 19, 2010.

Stout, David; Moore, Solomon (October 20, 2009). "U.S. won't prosecute in states that allow medical marijuana" (http://www.nytimes.com/2009/10/20/us/20cannabis.html). *The New York Times*: p. A1. . Retrieved July 19, 2010.

Johnson, Connie (October 20, 2009). "U.S. eases stance on medical marijuana. Attorney general says prosecuting such cases 'will not be a priority'" (http://www.washingtonpost.com/wp-dyn/content/article/2009/10/19/AR2009101903638.html). *The Washington Post*: p. A1. . Retrieved July 19, 2010.

[263] Egelko, Bob (December 18, 2009). "U.S. ends funding ban for needle exchanges" (http://www.sfgate.com/cgi-bin/article.cgi?f=/c/a/2009/12/18/MNKM1B5S7L.DTL). *San Francisco Chronicle*: p. A126. . Retrieved July 20, 2010.

[264] Jesse Lee. "The President Signs Repeal of "Don't Ask Don't Tell": "Out of Many, We Are One"" (http://www.whitehouse.gov/blog/2010/12/22/president-signs-repeal-dont-ask-dont-tell-out-many-we-are-one). Whitehouse.gov. . Retrieved December 22, 2010.

[265] "'Don't ask, don't tell' repealed as Obama signs landmark law" (http://www.guardian.co.uk/world/2010/dec/22/obama-repeals-dont-ask-dont-tell). *The Guardian* (London). December 22, 2010. . Retrieved December 22, 2010.

[266] "Obama to delay 'don't ask, don't tell' repeal" (http://www.washingtontimes.com/news/2008/nov/21/obama-to-delay-repeal-of-dont-ask-dont-tell/). *The Washington Times*. November 21, 2008. . Retrieved December 22, 2010.

[267] Robert Schroeder (5 November 2009). "Obama urges Congress to pass health-care reform" (http://articles.marketwatch.com/2009-11-05/news/30991733_1_health-care-reform-president-barack-obama-overhaul-bill). MarketWatch. . Retrieved 14 May 2012.

[268] U.S. Congressional Budget Office, *Preliminary Analysis of Major Provisions Related to Health Insurance Coverage Under the Affordable Health Choices Act* (http://www.cbo.gov/ftpdocs/103xx/doc10310/06-15-HealthChoicesAct.pdf), June 15, 2009

[269] Lori Montgomery and Shailagh Murray, "Lawmakers Warned About Health Costs: CBO Chief Says Democrats' Proposals Lack Necessary Controls on Spending," (http://www.washingtonpost.com/wp-dyn/content/article/2009/07/16/AR2009071602242.html) *The Washington Post*, July 17, 2009

[270] Reuters, "U.S. House Panel Passes Health Bill, Critics Slam Cost," (http://www.nytimes.com/reuters/2009/07/17/us/politics/politics-us-usa-healthcare-house.html) *The New York Times*, July 17, 2009

[271] "House Approves Health Overhaul, Sending Landmark Bill to Obama," (http://www.nytimes.com/2010/03/22/health/policy/22health.html) *The New York Times*, March 21, 2010

[272] "Obama presents health plan, attacks 'scare tactics'" - CNN.com (http://www.cnn.com/2009/POLITICS/09/09/obama.speech/index.html). Retrieved September 9, 2009.

[273] "Obama Not Worried About "Procedural Rules" like "Deem and Pass" for Health Care - Political Punch" (http://blogs.abcnews.com/politicalpunch/2010/03/obama-not-worried-about-procedural-rules-like-deem-and-pass-for-health-care.html). Blogs.abcnews.com. 2010-03-17. . Retrieved 2010-12-18.

[274] Somashekhar, Sandhya; Kane, Paul (March 18, 2010). "Democrats yet to decide on health-care bill bear the weight of Washington" (http://www.washingtonpost.com/wp-dyn/content/article/2010/03/17/AR2010031704184.html). *The Washington Post*. . Retrieved May 22, 2010.

[275] David Beardsley (25 March 2010). "Congress Passes Final Tweaking to Health Reform Bill" (http://www.wusa9.com/news/local/story.aspx?storyid=99102). . Retrieved 14 May 2012.

[276] Gregg Hitt (25 March 2010). "Congress Approves Final Health Overhaul" (http://online.wsj.com/article/SB10001424052748704094104575143603431943646.html). *The Wall Street Journal*. . Retrieved 14 May 2012.

[277] Herszenhorn, David M. (March 30, 2010). "The Last Piece in Place" (http://prescriptions.blogs.nytimes.com/2010/03/30/the-last-piece-in-place/). *The New York Times*. . Retrieved May 22, 2010.

[278] Montopoli, Brian (March 30, 2010). "Obama Lauds Passage of Education Reform" (http://www.cbsnews.com/8301-503544_162-20001419-503544.html). *CBS News*. .

[279] "What would change if student lending legislation passes" (http://www.washingtonpost.com/wp-dyn/content/article/2010/03/25/AR2010032503578.html). *The Washington Post*. March 26, 2010. . Retrieved May 22, 2010.

[280] Madison, Lucy (2011-10-26). "Who will benefit from Obama's student loan plan?" (http://www.cbsnews.com/8301-503544_162-20126172-503544/who-will-benefit-from-obamas-student-loan-plan/). CBS News. . Retrieved 2012-5-14.

[281] Associated Press (30 March 2010). "Q&A: What does student loan overhaul mean for U.S., you?" (http://www.usatoday.com/news/education/2010-03-30-student-loans_N.htm). *USA Today*. . *Retrieved 14 May 2012.*

[282] "Obama on 'milestone' education bill" (http://firstread.msnbc.msn.com/archive/2010/03/30/2246401.aspx). msnbc.com. 2010-03-30. . Retrieved 2010-12-18.

[283] Khan, Huma; McPhee, Michele; Goldman, Russell (July 24, 2009). "Obama Called Police Officer Who Arrested Gates, Still Sees 'Overreaction' in Arrest" (http://abcnews.go.com/Politics/story?id=8163051&page=1). . Retrieved 07-12-2010.

[284] Censky, Annalyn (July 21, 2010). "Obama on new law: 'No more taxpayer bailouts'" (http://money.cnn.com/2010/07/21/news/economy/obama_signs_wall_street_reform_bill/index.htm). *CNN*. . Retrieved 07-22-2010.

[285] Espo, David; Pace, Julie (November 3, 2010). "Obama Takes Responsibility For Voter Frustration On Election Night" (http://www.huffingtonpost.com/2010/11/03/obama-takes-responsibility_n_778302.html). *Huffington Post*. . Retrieved April 3, 2011.

[286] "GOP roars back to take U.S House; Dems cling to Senate Majority" (http://www.cnn.com/2010/POLITICS/11/02/election.main/index.html). *Election Center* (CNN). November 3, 2010. . Retrieved April 3, 2011.

[287] Richardson, Chris (November 4, 2010). "Obama calls midterm elections a 'shellacking' for Democrats" (http://www.csmonitor.com/USA/Elections/Vox-News/2010/1104/Obama-calls-midterm-elections-a-shellacking-for-Democrats). *The Christian Science Monitor*. . Retrieved April 3, 2011.

[288] Norris, Michele; Siegel, Robert (November 3, 2010). "Obama: Midterm Election Was A 'Shellacking'" (http://www.npr.org/templates/story/story.php?storyId=131048554). NPR. . Retrieved April 3, 2011.

[289] "Lack of change you can believe in" (http://www.economist.com/blogs/democracyinamerica/2011/04/barack_obamas_re-election_campaign). *The Economist*. .

External links

* **"President Barack Obama: The Man and His Journey"** (http://rcm.amazon.com/e/cm?t=monmakblo0a-20/
)

- "Obama's People" (http://www.nytimes.com/packages/html/magazine/2009-inauguration-gallery/index.html) (photography: Nadav Kander)
- The Official White House YouTube page (http://www.youtube.com/user/whitehouse)
- "President Barack Obama's Inaugural Address" (http://www.whitehouse.gov/blog/inaugural-address/). The White House.
- "Wrapping Up Open for Questions" (http://www.whitehouse.gov/blog/09/03/26/Wrapping-Up-Open-for-Questions/). The White House.
- "President Obama's State of the Union Address" (http://cspan.org/Events/President-Obamas-State-of-the-Union-Address/10737419121-6/) C-SPAN.

United_States_Secretary_of_Transportation

United States Secretary of Transportation

Official Seal

Incumbent
Ray LaHood
since January 23, 2009

Formation	October 15, 1966
First holder	Alan Stephenson Boyd
Succession	14th
Website	www.dot.gov [28]

The **United States Secretary of Transportation** is the head of the United States Department of Transportation, a member of the President's Cabinet, and fourteenth in the Presidential line of succession.[1] The post was created with the formation of the Department of Transportation on October 15, 1966, by President Lyndon B. Johnson's signing of the Department of Transportation Act.[2] The Department's mission is "to develop and coordinate policies that will provide an efficient and economical national transportation system, with due regard for need, the environment, and the national defense."[2] The Secretary of Transportation oversees eleven agencies, including the Federal Aviation Administration, the Federal Highway Administration, and the National Highway Traffic Safety Administration.[2] In April 2008, Mary Peters launched the official blog of the Secretary of Transportation called The Fast Lane.[3]

The first Secretary of Transportation was Alan Stephenson Boyd, nominated to the post by Democratic President Lyndon B. Johnson. Ronald Reagan's second Secretary of Transportation, Elizabeth Dole, was the first female holder, and Mary Peters was the second. Gerald Ford's nominee William Thaddeus Coleman, Jr. was the first African American to serve as Transportation Secretary, and Federico Peña, serving under Bill Clinton, was the first Hispanic to hold the position, subsequently becoming Secretary of Energy. Japanese American Norman Mineta, who had previously been Secretary of Commerce, is the longest-serving Secretary, holding the post for over five and a half years,[2] and Andrew Card is the shortest-serving Secretary, serving only eleven months. Neil Goldschmidt was the youngest secretary, taking office at age thirty-nine, while Norman Mineta was the oldest, retiring at age seventy-four.[4] On January 23, 2009, the sixteenth and current secretary Ray LaHood took office, serving under the administration of Democrat Barack Obama; he had previously been a Republican Congressman from Illinois for fourteen years.[5] The salary of the Secretary of Transportation is $199,700.[6]

Secretaries of Transportation

Parties

Democratic Republican

No.	Portrait	Name	State of Residence	Took Office	Left Office	President(s)
1		Alan Stephenson Boyd	Florida	January 16, 1967	January 20, 1969	Lyndon B. Johnson
2		John A. Volpe	Massachusetts	January 22, 1969	February 2, 1973	Richard Nixon
3		Claude Brinegar	California	February 2, 1973	February 1, 1975	Gerald Ford
4		William Thaddeus Coleman, Jr.	Pennsylvania	March 7, 1975	January 20, 1977	
5		Brock Adams	Washington	January 23, 1977	July 20, 1979	Jimmy Carter
6		Neil Goldschmidt	Oregon	August 15, 1979	January 20, 1981	
7		Drew Lewis	Pennsylvania	January 23, 1981	February 1, 1983	Ronald Reagan
8		Elizabeth Dole	Kansas	February 7, 1983	September 30, 1987	
9		James H. Burnley IV	North Carolina	December 3, 1987	January 30, 1989	

10		Samuel K. Skinner	Illinois	February 6, 1989	December 13, 1991	George H. W. Bush
11		Andrew Card	Massachusetts	February 24, 1992	January 20, 1993	
12		Federico Peña	Colorado	January 21, 1993	February 14, 1997	Bill Clinton
13		Rodney Slater	Arkansas	February 14, 1997	January 20, 2001	
14		Norman Mineta	California	January 25, 2001	August 7, 2006	George W. Bush
15		Mary Peters	Arizona	October 17, 2006	January 20, 2009	
16		Ray LaHood	Illinois	January 23, 2009	*Incumbent*	Barack Obama

Line of succession

The line of succession regarding who would act as Secretary of Transportation in the event of a vacancy or incapacitation is as follows:[7]

1. Deputy Secretary of Transportation
2. Under Secretary of Transportation for Policy
3. General Counsel
4. Assistant Secretary for Budget and Programs
5. Assistant Secretary for Transportation Policy
6. Assistant Secretary for Governmental Affairs.
7. Assistant Secretary for Aviation and International Affairs
8. Assistant Secretary for Administration
9. Administrator of the Federal Highway Administration
10. Administrator of the Federal Aviation Administration
11. Administrator of the Federal Motor Carrier Safety Administration
12. Administrator of the Federal Railroad Administration
13. Administrator of the Federal Transit Administration
14. Administrator of the Maritime Administration
15. Administrator of the Pipeline and Hazardous Materials Safety Administration
16. Administrator of the National Highway Traffic Safety Administration
17. Administrator of the Research and Innovative Technology Administration
18. Administrator of the Saint Lawrence Seaway Development Corporation
19. Regional Administrator, Southern Region, Federal Aviation Administration
20. Director, Resource Center, Lakewood, Colorado, Federal Highway Administration
21. Regional Administrator, Northwest Mountain Region, Federal Aviation Administration

References

General

- "Biographical Sketches of the Secretaries of Transportation" [8]. U.S. Department of Transportation. August 14, 2009. Retrieved January 2, 2010.

Specific

[1] "3 U.S.C. § 19 (http://www.law.cornell.edu/uscode/3/19.html) - Vacancy in offices of both President and Vice President; officers eligible to act". *Legal Information Institute*. Cornell University Law School.

[2] Grinder, R. Dale. "The United States Department of Transportation: A Brief History" (http://dotlibrary.dot.gov/Historian/history.htm). U.S. Department of Transportation. . Retrieved January 2, 2010.

[3] "A Chronology of Dates Significant in the Background, History and Development of the Department of Transportation" (http://dotlibrary.dot.gov/Historian/chronology.htm). U.S. Department of Transportation. August 14, 2009. . Retrieved January 3, 2010.

[4] "Biographical Sketches of the Secretaries of Transportation" (http://dotlibrary.dot.gov/Historian/bios.htm). U.S. Department of Transportation. August 14, 2009. . Retrieved January 3, 2010.

[5] "Ray LaHood—Secretary of Transportation" (http://www.dot.gov/bios/lahood.htm). U.S. Department of Transportation. July 22, 2009. . Retrieved January 3, 2010.

[6] "Salary Table No. 2010—Ex" (http://www.opm.gov/oca/10tables/html/ex.asp). U.S. Office of Personnel Management. . Retrieved January 3, 2010.
"5 U.S.C. § 5312 (http://www.law.cornell.edu/uscode/5/5312.html) - Positions at level I". *Legal Information Institute*. Cornell University Law School.

[7] Obama, Barack (January 14, 2009). "Executive Order 13485: Providing an Order of Succession Within the Department of Transportation" (http://nodis3.gsfc.nasa.gov/displayEO.cfm?id=EO_13485_). NASA Online Directives Information System. . Retrieved January 2, 2010.

[8] http://dotlibrary.dot.gov/Historian/bios.htm

The order of succession of the U.S. Department of Transportation is as follows (see http://edocket.access.gpo.gov/2008/pdf/E8-5543.pdf):

(1) Deputy Secretary. (2) Under Secretary of Transportation for Policy. (3) General Counsel. (4) Assistant Secretary for Budget and Programs. (5) Assistant Secretary for Transportation Policy. (6) Assistant Secretary for Governmental Affairs. (7) Assistant Secretary for Aviation and International Affairs. (8) Assistant Secretary for Administration. (9) Administrator of the Federal Highway Administration (10) Administrator of the Federal Aviation Administration (11) Administrator of the Federal Motor Carrier Safety Administration (12) Administrator of the Federal Railroad Administration (13) Administrator of the Federal Transit Administration (14) Administrator of the Maritime Administration (15) Administrator of the Pipeline and Hazardous Materials Safety Administration (16) Administrator of the National Highway Traffic Safety Administration (17) Administrator of the Research and Innovative Technology Administration (18) Administrator of the Saint Lawrence Seaway Development Corporation (19) Regional Administrator, Southern Region, Federal Aviation Administration (20) Director, Resource Center, Lakewood, Colorado, Federal Highway Administration (21) Regional Administrator, Northwest Mountain Region, Federal Aviation Administration [edit]

External links

- Department of Transportation (http://www.dot.gov/)
- The Fast Lane (http://fastlane.dot.gov/), the official blog of the Secretary

Ray_LaHood

Ray LaHood	
[[file:Ray LaHood official DOT portrait.jpg	alt=]]
16th United States Secretary of Transportation	
Incumbent	
Assumed office January 22, 2009	
President Barack Obama	
Deputy John Porcari	
Preceded by Mary Peters	
Member of the U.S. House of Representatives from Illinois's 18th district	
In office January 3, 1995 – January 3, 2009	
Preceded by Bob Michel	
Succeeded by Aaron Schock	
Member of the Illinois House of Representatives	
In office May 1982 – January 4, 1983	
Personal details	
Born Raymond H. LaHoodDecember 6, 1945Peoria, Illinois, United States	
Political party Republican	
Spouse(s) Kathy LaHood	
Alma mater Bradley University	
Profession Teacher[1]	
Religion Maronite Catholicism	

Raymond H. "Ray" LaHood (born December 6, 1945) is a Republican politician from Illinois who is currently the United States Secretary of Transportation, having served since 2009. Previously, he represented the Illinois's 18th congressional district in the U.S. House of Representatives for seven terms (1995–2009).

Early life and education

LaHood was born in Peoria, Illinois, to Edward M. LaHood, a Lebanese American who managed a restaurant, and Mary A. LaHood (née Vogel), who was of German ancestry.[2] [3] [4] He graduated from Spalding Institute, worked his way through Canton Junior College and Bradley University in Peoria, earning a Bachelors of Science in education and sociology in 1971.[2]

Early career

Following graduation, he taught junior high school social studies at public and Catholic schools,[2] and has said that "teaching kids ... about the constitution and government" stirred his interest in politics.[1]

LaHood was director of the Rock Island County Youth Services Bureau and then district administrative assistant for U.S. congressman Tom Railsback, a Moline, Illinois Republican, from 1977 to 1982.[5] He was appointed in 1982 to fill a vacant seat in the Illinois House of Representatives, serving for nine months, and running for the seat in November 1982, but losing to Democrat Bob DeJaegher.[5] LaHood then became administrative assistant and ultimately the chief of staff to U.S. House minority leader Robert Michel, serving from 1982 until 1994.[5]

U.S. Representative

When Michel announced his retirement in 1994, LaHood ran and won his seat in the House, representing Illinois's 18th congressional district.[2] LaHood was one of only three Republicans elected to the House that year who did not sign on to the Contract with America, Newt Gingrich's manifesto for a Republican majority,[5] [6] and was a member of the moderate Republican Main Street Partnership. In 1997, in an effort to promote bipartisan cooperation, LaHood organized bipartisan retreats for members of Congress.[5]

During his service in Congress, he became well-known among C-SPAN viewers as the presiding officer of more debates than any other member.[2] Most notably, in 1998 he presided over the contentious debate over the impeachment of President Bill Clinton.[5] [7]

A strong advocate for preserving the legacy of Abraham Lincoln, LaHood authored a law that established the Abraham Lincoln Bicentennial Commission, which laid the groundwork for celebrating the 16th President's 200th birthday in 2009. He has also been a lead Capitol Hill supporter for the Lincoln Presidential Library in Springfield, Illinois, and is one of 15 members on the ALBC.

LaHood served on the House Transportation and Infrastructure Committee from 1995 until 2000,[8] the House Permanent Select Committee on Intelligence beginning in 1998, and the House Appropriations Committee beginning in 2000.[5] In 2005 he voted against renewing the PATRIOT Act, saying he opposed extending its intrusive police powers.[5]

LaHood was said to be considering a challenge to Governor Rod Blagojevich's re-election bid in 2006, but chose to run for another term in Congress instead.[5] He won the 2006 race against Steve Waterworth[9] by a margin of 147,108 (67%) to 71,106 (33%).[10] On July 26, 2007, LaHood stated he would not seek re-election in 2008.[11]

In August 2007, LaHood received a 0% rating from the conservative and anti-earmark Club for Growth 2007 RePORK Card.[12] He received an 11% rating from the conservative lobbying group Citizens Against Government Waste in August 2007, and holds a lifetime 49% rating from the group.[13]

In 2007 LaHood considered, but later decided against, applying for the post of president of Bradley University.[5]

During the 2008 presidential election, LaHood supported John McCain, but criticized the rallies being held by McCain's vice presidential nominee, Sarah Palin, saying she should put a stop to the name calling, and that the tactic could backfire. "This doesn't befit the office that she's running for. And frankly, people don't like it," he said.[14] [15]

Secretary of Transportation

On December 19, 2008, President-elect Barack Obama announced that he would nominate Ray LaHood to be the next Transportation Secretary. LaHood's résumé on transport matters was considered thin by some critics, including the *Wall Street Journal* despite the fact that he served on the House Transportation and Infrastructure Committee.[8] As a member of the House Appropriations Committee he won praise for his "skills as an arbiter" in being able to bridge sometimes bitter partisan divides in the Congress, something the position would require.[16] Some critics alleged a reputation for pork barrel spending, including in support of campaign contributors. *The Washington Post* reported that of the $60 million in earmarks LaHood secured for his district in 2008, $9 million went to campaign donors.[17]

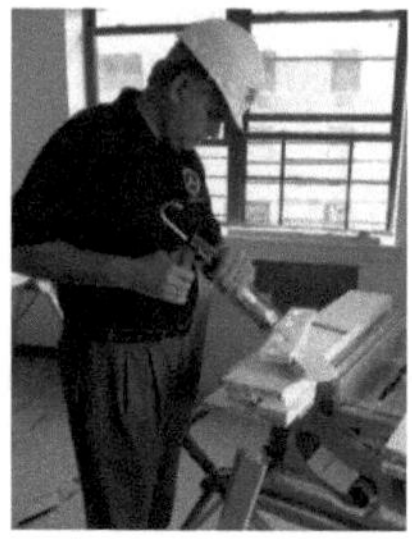

LaHood works on a Habitat for Humanity project in Brooklyn, New York City, June 2009

His nomination was confirmed by the Senate by voice vote on January 21, 2009.[18] He was, with Robert Gates,[19] [20] one of two Republican members of the Obama Cabinet.[21]

On February 3, 2010, LaHood was criticized for advice he was asked to give while testifying before a congressional committee regarding Toyota's recall of 2.3 million vehicles due to sudden acceleration, wherein he suggested Toyota owners stop driving their cars. LaHood qualified his statement within an hour and a half of his testimony, spelling out that he meant "owners of any recalled Toyota models (should) contact their local dealer and get their vehicles fixed as soon as possible."[22]

Ray LaHood is a supporter of airline passenger rights to facilities, food and water during lengthy on-aircraft delays.[23] He is also a strong proponent of high-speed rail, saying "This is what the American people want. If you build it, they will come."[24]

LaHood plans to step down as transportation secretary at the end of Obama's first term in 2013. He will not seek any public office after that, and instead enter the private sector.[25]

On December 6, 2011, Secretary LaHood accepted the resignation of FAA Administrator Randy Babbitt, who was charged with drunk driving near his Washington home.

Egypt

On January 21, 2012, LaHood's son, Sam LaHood, was detained by the Egyptian government and not allowed to leave the country as part of a politically charged criminal investigation by the Egyptian government into the activities of non-governmental organizations (NGOs) monitoring local elections in Egypt. LaHood's son is the Egypt director of the International Republican Institute. The Egyptian government has detained twelve NGO representatives from leaving Egypt.[26]

On February 5, 2012, Egyptian authorities charged LaHood's son and 42 other individuals with "spending money from organizations that were operating in Egypt without a license." Nineteen Americans are part of the 42 charged. The U.S. government has made it clear that $1.5 billion in U.S. aid to Egypt could be withheld if the investigation is not finished quickly. Faiza Abu Naga, Egypt's Minister of International Cooperation, is seen as the person pushing the investigation forward.[27] [28]

Electoral history

Ray LaHood's congressional seat, Illinois's 18th congressional district, has been Republican since 1939.

Illinois's 18 congressional district: Results 1994–2006[29]

Year	Democrat	Votes	Pct	Republican	Votes	Pct	
1994	G. Douglas Stephens	78,332	39%	**Ray LaHood**	119,838	60%	*
1996	Mike Curran	98,413	41%	**Ray LaHood**	143,110	59%	
1998	*(no candidate)*			**Ray LaHood**	158,175	100%	*
2000	Joyce Harant	85,317	33%	**Ray LaHood**	173,706	67%	
2002	*(no candidate)*			**Ray LaHood**	192,567	100%	
2004	Steve Waterworth	91,548	30%	**Ray LaHood**	216,047	70%	
2006	Steve Waterworth	73,052	33%	**Ray LaHood**	150,194	67%	

* Write-in and minor candidate notes: In 1994, write-ins received 955 votes. In 1998, write-ins received 2 votes.

LaHood did not to seek re-election in 2008, and Barack Obama nominated him to be U.S. Secretary of Transportation. Illinois State Representative Aaron Schock of Peoria won the seat for the Republicans in the 2008 election.[30]

Personal life

LaHood and his wife Kathy have a residence in East Peoria, Illinois. Ray and Kathy have four children--Darin, Amy, Sam, Sara. Their son, Darin LaHood, is a current member of the Illinois Senate. Their other son, Sam LaHood, is currently Director of the International Republican Institute office in Egypt, a nongovernmental organization assisting democratization efforts, which has come under investigation by the Egyptian government; he was denied permission to leave the country because he was told he was on a "no-fly list"[31] .

See also

- List of U.S. political appointments that crossed party lines

References

[1] "Rep. LaHood Talks History at Holy Family School" (http://www.week.com/news/local/30629004.html). Peoria, Illinois: WEEK-TV. October 8, 2008. . Retrieved February 5, 2011.

[2] U.S. Congressman Ray LaHood (Archived version from 2003) (http://web.archive.org/web/20041011111648/www.house.gov/lahood/biography2003.html)

[3] "Ancestries of Miscellaneous Celebrities, Ray LaHood" (http://freepages.genealogy.rootsweb.ancestry.com/~battle/celeb/lahood.htm). Ancestry.com. . Retrieved February 4, 2010.

[4] "Famous Arab Americans - Political" (http://www.aaiusa.org/pages/famous-arab-americans/#Politics). Arab American Institute. . Retrieved February 4, 2010.

[5] "Ray LaHood career highlights" (http://www.pjstar.com/news/x415866436/Career-highlights). *Peoria Journal Star*. December 6, 2008. . Retrieved February 10, 2011.

[6] Fletcher, Michael A.; Rucker, Philip (December 18, 2008). "Obama to Add GOP's LaHood to Cabinet" (http://www.washingtonpost.com/wp-dyn/content/article/2008/12/17/AR2008121703483.html?hpid=topnews). Washington Post. .

[7] =Broder, David S. (August 2, 2007). "A Setback for Civility" (http://www.washingtonpost.com/wp-dyn/content/article/2007/08/01/AR2007080102050.html). *The Washington Post*. . Retrieved February 5, 2011.

[8] Ray LaHood - U.S. Department of Transportation (http://www.dot.gov/bios/lahood.htm)

[9] Waterworth for Congress (http://www.waterworthforcongress.com/)

[10] "Race finder, Illinois' 18th Congressional District" (http://www.cnn.com/ELECTION/2006/pages/results/states/IL/H/18/district.html). *Elections 2006"* (CNN). .

[11] Tankersley, Jim; Pearson, Rick (July 26, 2007). "LaHood will not seek re-election" (http://www.swamppolitics.com/news/politics/blog/2007/07/lahood_will_not_seek_reelectio.html). *The Swamp* (Chicago Tribune). .

[12] The 2007 Club for Growth RePORK Card (http://www.clubforgrowth.org/2007/08/the_2007_club_for_growth_repor.php) "The Club For Growth", August 9, 2007

[13] Citizens Against Government Waste 2007 House Scorecard (http://councilfor.cagw.org/site/DocServer/2007_House_Ratings_Final.pdf?docID=3241) "Citizens Against Government Waste", August 27, 2008

[14] Nagle, Maria (October 10, 2008). "LaHood criticizes Palin campaign rallies". *Jacksonville Journal-Courier* (Jacksonville, Illinois).

[15] "Republican Congressman Criticizes Palin Rallies" (http://www.huffingtonpost.com/2008/10/10/republican-congressman-cr_n_133623.html). *The Huffington Post*. October 10, 2008. . Retrieved February 11, 2011.

[16] Weisman, Jonathan; Conkey, Christopher (December 18, 2008). "LaHood to Get Transportation Post" (http://online.wsj.com/article/SB122955244259215675.html). *Wall Street Journal*. . Retrieved December 18, 2008.

[17] Leonnig, Carol D. (January 14, 2009). "Last Year, LaHood Sponsored Millions in Earmarks; Some Funding Went to Campaign Donors" (http://www.washingtonpost.com/wp-dyn/content/article/2009/01/13/AR2009011302860.html). *The Washington Post*. . Retrieved May 22, 2010.

[18] Matthew L. Wald (January 22, 2009). "Panel Approves Transportation Nominee" (http://www.nytimes.com/2009/01/22/us/politics/22lahood.html). *New York Times*. .

[19] Gates is not registered with any political party, but considers himself Republican. "Gates: Military looks to accelerate Iraq pullout" (http://www.msnbc.msn.com/id/28022197/). Associated Press. Associated Press. December 1, 2008. . Retrieved May 5, 2009.

[20] Wall Street Journal "bipartisan" (http://online.wsj.com/article/SB122955244259215675.html)

[21] Wald, Matthew L. (January 22, 2009). "Panel Approves Transportation Nominee" (http://www.nytimes.com/2009/01/22/us/politics/22lahood.html). *The New York Times*. . Retrieved May 22, 2010.

[22] "Toyota recall:DOT Secretary LaHood pulls back from telling owners not to drive their cars" (http://content.usatoday.com/communities/driveon/post/2010/02/toyota-recall-transportation-secretary-ray-lahood-advises-owners-not-to-drive-their-cars/1). USA Today. February 3, 2010. . Retrieved February 5, 2011.

[23] Maxon, Terry (May 12, 2010). "DOT Secretary LaHood names aviation future panel" (http://aviationblog.dallasnews.com/archives/2010/05/dot-secretary-lahood-names-avi.html). *Aviation Blog* (DallasNews.com). . Retrieved February 7, 2011.

[24] Chapman Steve (2011-04-28) Taking Taxpayers for a Ride (http://reason.com/archives/2011/04/28/taking-taxpayers-for-a-ride/singlepage), *Reason*

[25] Skiba, Katherine (October 13, 2011), "LaHood to be transportation secretary for one term only" (http://www.chicagotribune.com/news/local/breaking/chi-lahood-to-be-transportation-secretary-for-one-term-only-20111013,0,1233656.story), *Chicago Tribune* (Chicago),

[26] Steven Myers and David Kirkpatrick (January 26, 2012). "As Tensions Rise, Egypt Bars Exit of Six Americans" (http://www.nytimes.com/2012/01/27/world/middleeast/egypt-bars-son-of-ray-lahood-from-leaving.html). *New York Times*. . Retrieved 26 January 2012.

[27] Londoño, Ernesto and William Wan (5 February 2012). "Egypt to prosecute Americans in NGO probe; Sam LaHood among those facing criminal charges" (http://www.washingtonpost.com/world/egypt-to-prosecute-americans-in-ngo-probe/2012/02/05/gIQAQRderQ_story.html?tid=pm_pop). The Washington Post. . Retrieved 7 February 2012.

[28] Bradley, Matt (February 7, 2012). "Egypt Names 19 Americans in Probe" (http://online.wsj.com/article/SB10001424052970203315804577206922611222662.html?mod=ITP_pageone_3). . Retrieved 7 February 2012.

[29] "Election Statistics" (http://clerk.house.gov/member_info/electionInfo/index.html). Office of the Clerk of the House of Representatives. . Retrieved January 10, 2008.

[30] "Elections" (http://www.week.com/news/elections). *WEEK News 25 website*. Granite Broadcasting. November 5, 2008. . Retrieved November 5, 2008. 100% of precincts reporting. Unframed data at (http://bimedia.ftp.clickability.com/weekwebftp/election/WEBRACES.HTM).

[31] Bresahan, John; Rogers, David (January 26, 2012), "Ray LaHood's son barred from leaving Egypt" (http://www.politico.com/news/stories/0112/71981.html), *POLITICO* (Arlington, VA),

External links

- Secretary Ray LaHood (http://www.dot.gov/bios/lahood.htm) at the United States Department of Transportation
- Ray LaHood on the Daily Show, December 15, 2009 (http://www.thedailyshow.com/watch/tue-december-15-2009/ray-lahood)
- Biography (http://www.whorunsgov.com/Profiles/Raymond_H._LaHood) at *WhoRunsGov.com* at *The Washington Post*
- Biography (http://bioguide.congress.gov/scripts/biodisplay.pl?index=l000552) at the *Biographical Directory of the United States Congress*
- Voting record (http://projects.washingtonpost.com/congress/members/l000552) maintained by *The Washington Post*

- Biography (http://www.votesmart.org/candidate/21699), voting record (http://www.votesmart.org/candidate/key-votes/21699), and interest group ratings (http://www.votesmart.org/candidate/evaluations/21699) at Project Vote Smart
- Congressional profile (http://www.govtrack.us/congress/person.xpd?id=400228) at GovTrack
- Issue positions and quotes (http://www.ontheissues.org/IL/Ray_LaHood.htm) at On the Issues
- Current Bills Sponsored (http://www.statesurge.com/members/839) at StateSurge.com
- Financial information (http://www.opensecrets.org/politicians/summary.php?CID=N00004933) at OpenSecrets.org
- Staff salaries, trips and personal finance (http://www.legistorm.com/member/326/Rep_Ray_LaHood.html) at LegiStorm.com
- Campaign finance reports and data (http://herndon1.sdrdc.com/cgi-bin/can_detail/H4IL18059) at the Federal Election Commission
- Appearances (http://www.c-spanvideo.org/person/26394) on C-SPAN programs
- Collected news and commentary (http://topics.nytimes.com/top/reference/timestopics/people/l/ray_lahood) at *The New York Times*
- C-SPAN *Q&A* interview with LaHood, September 2, 2007 (http://www.q-and-a.org/Program/?ProgramID=1142)
- Profile (http://www.sourcewatch.org/index.php?title=Ray_LaHood) at SourceWatch
- Works by or about Ray LaHood (http://worldcat.org/identities/lccn-no2004-3058) in libraries (WorldCat catalog)

Articles

- LaHood tapped for Obama Cabinet (http://www.pjstar.com/news_state/x415883215/LaHood-tapped-for-Obama-Cabinet) Karen McDonald, *Peoria Journal Star*, December 17, 2008
- The GOP Goes South (http://www.washingtonpost.com/wp-dyn/content/article/2008/12/26/AR2008122601129.html?hpid=opinionsbox1) David S. Broder, "Washington Post," December 28, 2008, on Ray LaHood's selection for the Obama cabinet and what that means for the Republican Party
- Ray LaHood: A driving force in White House (http://www.politico.com/news/stories/0511/54242.html)

Federation

A **federation** (Latin: *foedus, foederis*, 'covenant'), also known as a **federal state**, is a type of sovereign state characterized by a union of partially self-governing states or regions united by a central (federal) government. In a federation, the self-governing status of the component states is typically constitutionally entrenched and may not be altered by a unilateral decision of the central government.

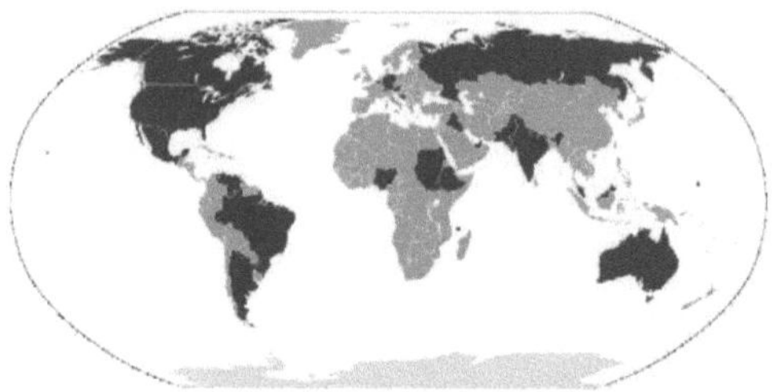

A map displaying current official federations

The form of government or constitutional structure found in a federation is known as **federalism**. It can be considered the opposite of another system, the unitary state. Germany with sixteen *Länder* is an example of a federation, whereas neighboring Austria and its *Bundesländer* was a unitary state with administrative divisions that became federated, and neighboring France by contrast has always been unitary.

Federations may be multi-ethnic and cover a large area of territory (eg.India), although neither is necessarily the case. The initial agreements create a stability that encourages other common interests, brings the disparate territories closer, and gives them all even more common ground. At some time this is recognized and a movement is organized to merge more closely. Other times, especially when common cultural factors are at play such as ethnicity and language, some of these steps in this pattern are expedited and compressed.

The international council for federal countries, the Forum of Federations,[1] is based in Ottawa, Canada. It helps share best practices among countries with federal systems of government, and currently includes nine countries as partner governments.

History

Several ancient chiefdoms and kingdoms, such as The 4th century BC League of Corinth, Noricum in Central Europe, and the Iroquois in pre-Columbian North America, could be described as federations or confederations. The Old Swiss Confederacy was an early example of formal non-unitary statehood.

Several colonies and dominions in the New World consisted of autonomous provinces, transformed to federal states upon independence (see Spanish American wars of independence). The oldest continuous federation, and a role model for many subsequent federations, is the United States of America. Some of the New World federations failed; the Federal Republic of Central America broke up into independent states 20 years after its founding. Others, such as Argentina and Mexico, have shifted between federal, confederal and unitary systems, before settling as federations. Brazil became a federation only after fall of the monarchy (see States of Brazil), and Venezuela became a federation after the Federal War.

Germany is another nation-state that has switched between confederal, federal and unitary rule, since the German Confederation was founded in 1815. The North German Confederation, the Weimar Republic and West Germany were federations.

Founded in 1922, the Soviet Union was formally a federation of Soviet Republics, Autonomous republics of the Soviet Union and other federal subjects, though in practice highly centralized under the Government of the Soviet Union. The Russian Federation has inherited a similar system.

Several dominons of the British Empire, independent during the years past World War II, became federations: Nigeria, Pakistan, India and Malaysia. Australia and Canada are independent federations, yet Commonwealth realms.

The Forum of Federations was established in 1999.

In some recent cases, federations have been a measure to handle with ethnic conflict within a state, such as Bosnia and Hercegovina and Iraq since 2005.

Federations and other forms of state

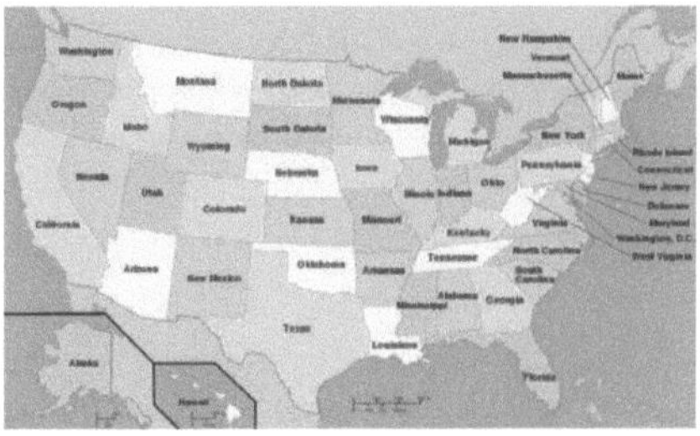

A map of the United States of America showing its fifty constituent states and the Federal District

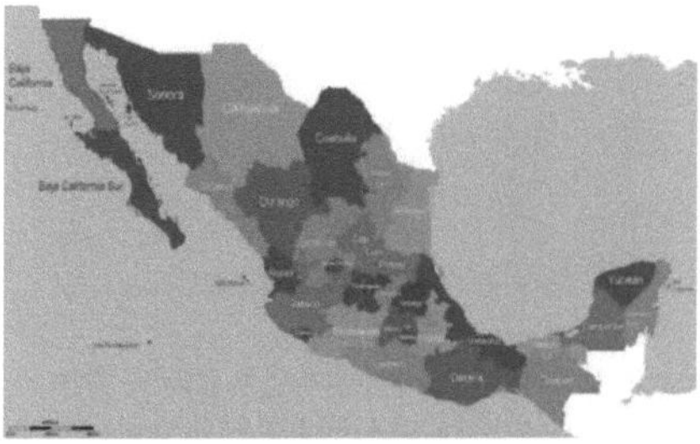

A map of the United Mexican States (Mexico), showing its thirty-one constituent states and the Federal District

Federations

In a federation the component states are in some sense sovereign, insofar as certain powers are reserved to them that may not be exercised by the central government. However, a federation is more than a mere loose alliance of independent states. The component states of a federation usually possess no powers in relation to foreign policy, and so they enjoy no independent status under international law. However, German Lander do have this power,[2] which is beginning to be exercised on a European level.

Some federations are called asymmetric because some states have more autonomy than others. An example of such a federation is Malaysia, in which Sarawak and Sabah entered the federation on different terms and conditions from the states of Peninsular Malaysia.

A federation often emerges from an initial agreement between a number of separate states. The purpose can be the will to solve mutual problems and to provide for mutual defense, or to create a nation state for an ethnicity spread over several states. The former was the case with the United States and Switzerland, the latter with Germany. However, as the histories of countries and nations vary, the federalist system of a state can be quite different from these models. Australia, for instance, is unique in that it came into existence as a nation by the democratic vote of the citizens of each state, who voted "yes" in referendums to adopt the Australian Constitution. Brazil, on the other hand, has experienced both the federal and the unitary state through its history. Some present day states of the Brazilian federation retain borders set during the Portuguese colonization (i.e. previous to the very existence of Brazilian state), whereas the latest state, Tocantins, was created by the 1988 Constitution for chiefly administrative reasons.

Seven of the top ten largest countries by area are governed as federations.

Unitary states

A unitary state is sometimes one with only a single, centralised, national tier of government. However, unitary states often also include one or more self-governing regions. The difference between a federation and this kind of unitary state is that in a unitary state the autonomous status of self-governing regions exists by the sufferance of the central government, and may be unilaterally revoked. While it is common for a federation to be brought into being by agreement between a number of formally independent states, in a unitary state self-governing regions are often created through a process of devolution, where a formerly centralised state agrees to grant autonomy to a region that was previously entirely subordinate. Thus federations are often established voluntarily from 'below' whereas devolution grants self-government from 'above'.

It is often part of the philosophy of a unitary state that, regardless of the actual status of any of its parts, its entire territory constitutes a single sovereign entity or nation-state, and that by virtue of this the central government exercises sovereignty over the whole territory *as of right*. In a federation, on the other hand, sovereignty is often regarded as residing notionally in the component states, or as being shared between these states and the central government.

Other forms of governance

Confederation

A confederation, in modern political terms, is usually limited to a permanent union of sovereign states for common action in relation to other states.[3]

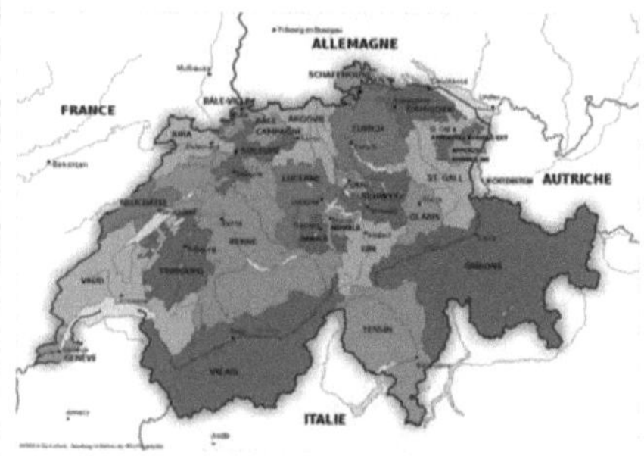

The Swiss Confederation and its 26 cantons

In Belgium, however, the opposite movement is under way.[4] Belgium was founded as a centralised state, after the French model, but has gradually been reformed into a federal state by consecutive constitutional reforms since the 1970s. Moreover, although nominally called a federal state, the country's structure already has a number of confederational traits (ex. competences are exclusive for either the federal or the state level, the treaty-making power of the Federating units without almost any possible veto of the Federal Government). At present, there is a growing movement to transform the existing federal state into a looser confederation with two or three constitutive states and/or two special regions.[5]

By definition, the difference between a confederation and a federation is that the membership of the member states in a confederation is voluntary, while the membership in a federation is not. A confederation is most likely to feature these differences over a federation: (1) No real direct powers: many confederal decisions are externalised by member-state legislation. (2) Decisions on day-to-day-matters are not taken by simple majority but by special majorities or even by consensus or unanimity (veto for every member). (3) Changes of the constitution, usually a treaty, require unanimity.

Over time these terms acquired distinct connotations leading to the present difference in definition. An example of this is the United States under the Articles of Confederation. The Articles established a national government under what today would be defined as a federal system (albeit with a comparatively weaker federal government). However, Canadians, designed with a *stronger* central government than the U.S. in the wake of the Civil War of the latter, use the term "Confederation" to refer to the formation or joining, not the structure, of Canada. Legal reforms, court rulings, and political compromises have somewhat *decentralised* Canada in practice since its formation in 1867.

Empire

An empire is a multi-ethnic state or group of nations with a central government established usually through coercion (on the model of the Roman Empire). An empire often includes self-governing regions, but these will possess autonomy only at the sufferance of the central government. On the other hand, a political entity that is an empire in name, may in practice consist of multiple autonomous kingdoms organised together in a federation, with a high king designated as an emperor. One example of this was Imperial Germany.

Comparison to other systems of autonomy

Federacy

A federacy is essentially an extreme case of an asymmetric federation, either due to large differences in the level of autonomy, or the rigidity of the constitutional arrangements. The term federacy is more often used for the relation between the sovereign state and its autonomous areas.

Devolution

A federation differs from a devolved state, such as the United Kingdom and the Kingdom of Spain, because, in a devolved state, the central government can revoke the independence of the subunits (Scottish Parliament, Welsh National Assembly, Northern Ireland Assembly in the case of the UK) without changing the constitution.

Associated States

A federation also differs from an associated state, such as the Federated States of Micronesia (in free association with the United States) and Cook Islands and Niue (which form part of the Realm of New Zealand). There are two kinds of associated states: in case of Micronesia, association is concluded by treaty between two sovereign states; in case of Cook Islands and Niue, association is concluded by domestic legal arrangements.

Crown dependencies

The relation between the Crown dependencies of the Isle of Man and the bailiwicks of Guernsey and Jersey in the Channel Islands and the United Kingdom is very similar to a federate relation: the Islands enjoy independence from the United Kingdom, which, via The Crown, takes care of their foreign relations and defence – although the UK Parliament does have overall power to legislate for the dependencies. However, the islands are neither an incorporated part of the United Kingdom, nor are they considered to be independent or associated states. The Isle of Man does not have a monarch, *per se*; rather, the British Monarch is, *ex officio*, Lord of Mann (irrespective of the incumbent's sex).

Overseas territories

Overseas territories, such as the British overseas territories, are vested with varying degrees of power; some enjoy considerable independence from the sovereign state, which only takes care of their foreign relations and defence. However, they are neither considered to be part of it, nor recognised as sovereign or associated states.

Alleged *de facto* federations

The distinction between a federation and a unitary state is often quite ambiguous. A unitary state may closely resemble a federation in structure and, while a central government may possess the theoretical right to revoke the autonomy of a self-governing region, it may be politically difficult for it to do so in practice. The self-governing regions of some unitary states also often enjoy greater autonomy than those of some federations. For these reasons, it is sometimes argued that some modern unitary states are *de facto* federations.

Spain

Autonomous communities of Spain

Spain is suggested as one possible *de facto* federation as it grants more self-government to its autonomous communities[6] [7] than most federations allow their constituent parts. For the Spanish parliament to revoke the autonomy of regions such as Galicia, Catalonia or the Basque Country would be a political near-impossibility, though nothing bars it legally. Additionally, some regions such as Navarre or the Basque Country have full control over taxation and spending, transferring a small payment to the central government for the common services (army, foreign relations, macroeconomic policy). For example, one scholar discusses the "federal nature of Spain's government (a trend that almost no one denies)."[8] Each autonomous community is governed by a Statute of Autonomy (*Estatuto de Autonomía*) under the Spanish Constitution of 1978.

People's Republic of China

In the People's Republic of China, a form of de facto federation has evolved without formal legislation. This has occurred as largely informal grants of power to the provinces, to handle economic affairs and implement national policies. This has resulted in a system some have termed "de facto federalism with Chinese characteristics" (in reference to Deng Xiaoping's policy of socialism with Chinese characteristics).[9] Constitutionally, the power vested in the special administrative regions of the People's Republic is granted from the Central People's Government, through decision by the National People's Congress. A Federal Republic of China, in effect the third Chinese Republic, has been proposed as a future replacement for the PRC.

European Union

The European Union (EU) is a type of political union or confederation. Robert Schuman, the initiator of the European Community system, wrote that a supranational Community like the Europe's founding European Coal and Steel Community lay midway between an association of States where they retained complete independence and a federation leading to a fusion of States in a super-state.[10] The European Founding Fathers made a Europe Declaration at the time of the signing of the Treaty of Paris on 18 April 1951 saying that Europe should be organized on a supranational foundation. They envisaged a structure quite different from a federation called the European Political Community.

The EU is a three pillar structure of the original supranational European Economic Community and the nuclear non-proliferation treaty, Euratom, plus two largely intergovernmental pillars dealing with External Affairs and Justice and Home Affairs. The EU is therefore not a *de jure* federation, although some academic observers conclude that after 50 years of institutional evolution since the Treaties of Rome it is *becoming* one.[11] The European Union possesses attributes of a federal state. However, its central government is far weaker than that of most federations and the individual members are sovereign states under international law, so it is usually characterized as an unprecedented form of supra-national union. The EU has responsibility for important areas such as trade, monetary union, agriculture, fisheries. Nonetheless, EU member states retain the right to act independently in matters of foreign policy and defense, and also enjoy a near monopoly over other major policy areas such as criminal justice and taxation. Since the Treaty of Lisbon, Member States' right to leave the Union is codified, and the Union operates with more qualified majority voting (rather than unanimity) in many areas.

By the signature of this Treaty, the participating Parties give proof of their determination to create the first supranational institution and that thus they are laying the true foundation of an organized Europe. This Europe remains open to all nations. We profoundly hope that other nations will join us in our common endeavour.

— **Europe Declaration** signed by Konrad Adenauer (West Germany), Paul van Zeeland, Joseph Meurice (Belgium) Robert Schuman (France) Count Sforza (Italy) Joseph Bech (Luxembourg) and Dirk Stikker, J. R. M. van den Brink (The Netherlands).[12]

Europe has charted its own brand of constitutional federalism.

— Joseph H. H. Weiler

Those uncomfortable using the "F" word in the EU context should feel free to refer to it as a quasi-federal or federal-like system. Nevertheless, for the purposes of the analysis here, the EU has the necessary attributes of a federal system. It is striking that while many scholars of the EU continue to resist analyzing it as a federation, most contemporary students of federalism view the EU as a federal system.
(*See for instance, Bednar, Filippov et al., McKay, Kelemen, Defigueido and Weingast*)

— R. Daniel Kelemen

A more nuanced view has been given by the German Constitutional Court. Here the EU is defined as 'an association of sovereign national states (*Staatenverbund*)'.[13] With this view, the European Union resembles more of a confederation.

Russian Federation

The Russian Federation has inherited its structure from the Russian Soviet Federative Socialist Republic (RSFSR) that was one of the 15 republics of the Soviet Union and itself was considered a federation of national territories. The RSFSR consisted of autonomous republics, which had a certain degree of autonomy, at least de jure, and of other types of administrative units (*oblasts* and *krais*), whose status was the same as that of oblasts in other – mostly unitary – Soviet Socialist Republics. Today's Russia is defined as a federation in its Constitution (Article 5),[14] and Russia's federal subjects, i.e., the constituent republics, oblasts, krais, the federal-level cities of Moscow and Saint Petersburg, as well as one autonomous oblast and four autonomous (national) *okrugs*, are equal in legal terms, save for some symbolic features allowed to republics (constitution, president, national language). Some regions (Yakutia[15] [16]) have concluded agreements with the Federation so as to modify the degree of their autonomy.

According to an amendment passed in December 2004, governors and presidents of Russia's constituent regions, who were previously elected by popular vote, are now proposed by the President of Russia for the approval of the local parliament[17] Local parliaments theoretically have the authority to reject the candidate, but if this occurs three times, the parliament may be dissolved by the President and new parliamentary elections held.

South Africa

At the end of apartheid, the former four provinces (Cape, Natal, Orange Free State and Transvaal) and the Bantustans were converted into nine provinces. Each province has a government constituted by a unicameral assembly and an executive branch with a premier and executive council.[18]

The powers of provinces are circumscribed by the national constitution which limits them to certain listed "functional areas". In certain cases there is a concurrence between national and provincial levels, while in others there are exclusive competences for each one. This system is known as "co-operative government".

This system would be considered as a "de facto" federation, because actually the power of provinces was given by the national government to the provincial level. Moreover, the creation of a "provincial constitution" is optional, if the province government does not want do it rules additionally the national constitution.

Ultimately, unlike the majority of federal systems, South Africa has a single national court system, and the administration of justice is the responsibility of the national government.

Internal controversy and conflict

Certain forms of political and constitutional dispute are common to federations. One issue is that the exact division of power and responsibility between federal and regional governments is often a source of controversy. Often, as is the case with the United States, such conflicts are resolved through the judicial system, which delimits the powers of federal and local governments. The relationship between federal and local courts varies from nation to nation and can be a controversial and complex issue in itself.

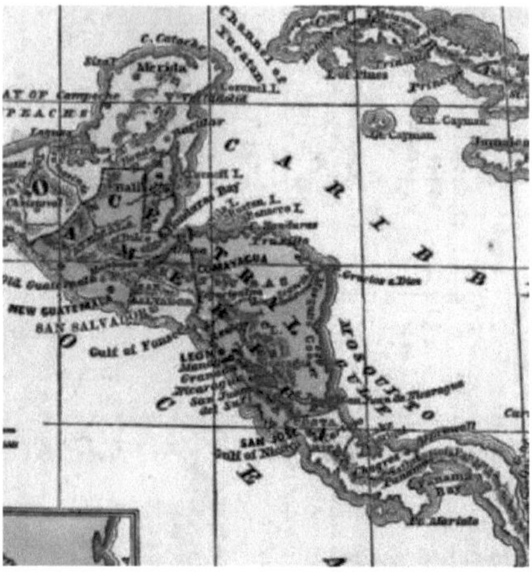

The United Provinces of Central America was short-lived

Another common issue in federal systems is the conflict between regional and national interests, or between the interests and aspirations of different ethnic groups. In some federations the entire jurisdiction is relatively homogeneous and each constituent state resembles a miniature version of the whole; this is known as 'congruent federalism'. On the other hand, incongruent federalism exists where different states or regions possess distinct ethnic groups.

The ability of a federal government to create national institutions that can mediate differences that arise because of linguistic, ethnic, religious, or other regional differences is an important challenge. The inability to meet this challenge may lead to the secession of parts of a federation or to civil war, as occurred in United States and Switzerland. In the case of Malaysia, Singapore was expelled from the federation because of rising racial tension. In some cases internal conflict may lead a federation to collapse entirely, as occurred in Nigeria, the Federation of Rhodesia and Nyasaland, the United Provinces of Central America and the West Indies Federation. Somalia, despite its Transitional Federal Charter, Transitional Federal Parliament, and Transitional Federal Government, has a weak central government: since 1992, federal institutions have rarely controlled territory outside of parts of Mogadishu (though this is changing) and the Somaliland region in the northwestern part of the country is autonomous.

Federal governments

The **federal government** is the common or national government of a federation. A federal government may have distinct powers at various levels authorized or delegated to it by its member states. The structure of federal governments vary. Based on a broad definition of a basic federalism, there are two or more levels of government that exist within an established territory and govern through common institutions with overlapping or shared powers as prescribed by a constitution.

Federal government is the government at the level of the sovereign state. Usual responsibilities of this level of government are maintaining national security and exercising international diplomacy, including the right to sign binding treaties. Basically, a modern federal government, within the limits defined by its constitution, has the power to make laws for the whole country, unlike local governments. The United States Constitution was created to limit the federal government from exerting power over the states, but certain amendments gave the federal government considerable authority over states.

Federal government within this structure are the government ministries and departments and agencies to which the ministers of government are assigned.

For a detailed list of federated units, see Federated state#List of constituents by federation. There are 27 federations as of July 2011.[19]

Contemporary

Year Est.	Federation	Federating Units	Major Federating Units	Minor Federating Units
	Argentina	Provinces of Argentina	23 provinces	1 autonomous city
1901	Australia	States and territories of Australia	6 states	10 territories
	Austria	States of Austria	9 *Länder* or *Bundesländer*	
	Belgium	Divisions of Belgium	3 Communities, 3 Regions	
	Bosnia and Herzegovina	Divisions of Bosnia and Herzegovina	2 entities (out of which one is itself a federation, consisting of 10 cantons)	1 district
1825	Brazil	States of Brazil	26 states (Previously provinces during the monarchical period between 1822 and 1889)	1 federal district and 5,561 municipalities
1867	Canada	Provinces and territories of Canada	10 provinces	3 territories
	Comoros		3 islands	
	Ethiopia	Regions of Ethiopia	9 regions	2 chartered cities
1949	Germany	States of Germany	16 *Länder* or *Bundesländer*	
1950	India	States and territories of India	28 States	7 Union Territories, including a National Capital Territory
	Iraq	Governorates of Iraq	18 provinces, including the sub-region of Iraqi Kurdistan.	
	Malaysia	States of Malaysia	13 states	3 federal territories
1821	Mexico	States of Mexico	31 states	1 federal district
	Federated States of Micronesia	Administrative divisions of the Federated States of Micronesia	4 states	
	Nepal	Zones of Nepal	14 zones	75 districts
	Nigeria	States of Nigeria	36 states	1 territory
	Pakistan	Provinces and territories of Pakistan	4 provinces	4 federal territories including a federal capital territory
1992	Russian Federation	Federal subjects of Russia	21 republics, 46 *oblasts*, 9 *krais*, 1 autonomous *oblast*, 4 autonomous *okrugs*, 2 federal-level cities[20]	
	Saint Kitts and Nevis	Islands/parishes of Saint Kitts and Nevis	2 islands/14 parishes	
1910	South Africa	Provinces of South Africa	9 provinces	
2011	South Sudan	States of South Sudan	10 states	
	Spain	Autonomous communities of Spain	17 autonomous communities	2 autonomous cities

	Sudan	States of Sudan	17 states	
	Switzerland	Cantons of Switzerland	20 cantons, 6 sous-cantons	
1973	United Arab Emirates	Emirates of the UAE	7 emirates	
1787	United States of America	States of the United States	50 states	1 federal district; 1 incorporated territory, 13 unincorporated territories
	Venezuela	States of Venezuela	23 states	1 federal district, 1 federal dependency

Long form titles

- ***Federal Republic*** of: Germany, Nigeria.
- ***Federation***: Russia, Saint Kitts and Nevis
- ***Federative Republic*** of: Brazil
- ***Republic*** of: Argentina, Austria, India (also called Indian Union), Iraq, South Africa, Sudan.
- **Others:**
  - *Bolivarian Republic* (Venezuela)
  - *Confederation* (Switzerland)
  - *Commonwealth* (Australia)
  - *Dominion* (Canada before 1982)
  - *Federal Democratic Republic* (Ethiopia, Nepal)
  - *Federated States* (FS Micronesia)
  - *Federative Republic* (Brazil)
  - *Islamic Republic* (Pakistan)
  - *Kingdom* (Belgium)
  - *Union* (Comoros)
  - *United Emirates* (United Arab Emirates)
  - *United Mexican States* (Mexico)
  - *United States of America* (United States)
- **None:**
  - Bosnia and Herzegovina (since 1997)
  - Canada (since 1982)
  - Malaysia

Defunct

- Federal Republic of Cameroon (1961–1972)
- United Provinces of Central America (1823 – circa 1838)
- Confederate States of America (1861–1865)
- United States of Colombia (1863–1886)
- Czechoslovakia (1969–1992)
- Federated Dutch Republic (1581–1795)
- French Equatorial Africa (1910–1960)
- French West Africa (1904–1958)
- Inca Empire (1197–1572)
- United States of Indonesia (1949–1950)
- United Kingdom of Libya (1951–1963)

- Federated Malay States (1896–1946)
- Federation of Malaya (1948–1963)[21]
- Mali Federation (1959–1960)
- New Granada (1855–1886)
- Polish-Lithuanian Commonwealth (1569–1795)
- Federation of Rhodesia and Nyasaland (1953–1963)
- Federation of South Arabia (1962–1967)
- Union of Soviet Socialist Republics[22] (1922–1991)
- Federal Republic of Spain (1873–1874)
- Uganda (1962–1967)
- West Indies Federation (1958–1962)
- Socialist Federal Republic of Yugoslavia[23] (1943–1992)
- Federal Republic of Yugoslavia (1992–2003)

Some of the proclaimed Arab federations were confederations *de facto*.

See also

- Anti-Federalism, an 18th century movement in the United States
- Capital city
- Confederation
- Corporative federalism
- Constitutional economics
- Political economy
- Rule according to higher law
- European Union
- European Coal and Steel Community
- Federacy
- Federalism in Australia
- Federalist
- The Federalist Papers
- Federal monarchy
- Federated state
- Federation of Australia
- Foederati
- Indian Union
- International organisation
- Międzymorze (Intermarum)
- Multinational state
- Neo-functionalism
- New federalism
- Regional state
- Supranationalism
- Supranational union
- Unitary state
- United Federation of Planets
- World Federalist Movement
- Centre for Studies on Federalism

Footnotes

[1] Forum of Federations (http://www.forumfed.org)

[2] http://www.tandfonline.com/doi/abs/10.1080/09644009208404305

[3] Oxford English Dictionary

[4] One of the most important recent books about the Belgian institutions, written by one of the leading French-speaking jurists concludes : *Vers le confédéralisme* (Toward a Confederation). See: Charles-Etienne Lagasse, *Les nouvelles institutions politiques de la Belgique et de l'Europe*, Erasme, Namur 2003, p. 603 ISBN 2-87127-783-4

[5] Many Flemings would prefer two states, Flanders and Wallonia, and two special regions, Brussels and the German-speaking region. In Wallonia, there is a wider support for three states : Flanders, Wallonia and Brussels.

[6] Mallet, Victor (18 August 2010). "Flimsier footings" (http://www.ft.com/cms/s/0/de6c00f0-8c25-11de-b14f-00144feabdc0.html). *Financial Times*. . Retrieved 25 August 2010.(**registration required**)

[7] "A survey of Spain: How much is enough?" (http://www.economist.com/specialreports/displaystory.cfm?story_id=12501023). The Economist. 6 November 2008. . Retrieved 25 August 2010.(**subscription required**)

[8] Enrique Guillén López (http://www.ugr.es/~pwdconst/enrique/ENRIQUE GUILLEN.htm), JUDICIAL REVIEW IN SPAIN: THE CONSTITUTIONAL COURT, 41 Loyola of Los Angeles Law Review 541, 544 (2008).

[9] Economic Warlords by Gregory H. Fuller (http://papers.ssrn.com/sol3/papers.cfm?abstract_id=1151187)

[10] *La Communaute du Charbon et de l'Acier*, p7 Paul Reuter with preface by Robert Schuman. Paris 1953.

[11] How the court made a federation of the EU. (http://www.springerlink.com/content/f3vw032557232l64/) Josselin (U de Rennes-1/CREM) and Marciano (U de Reims CA/CNRS).

[12] *Schuman or Monnet? The real Architect of Europe.* p 129. Bron 2004

[13] *BVerfG, 2 BvE 2/08 vom 30.6.2009, Absatz-Nr. (1 - 421)* (http://www.bverfg.de/entscheidungen/es20090630_2bve000208en.html)

[14] Constitutional definition of Russia as a federation (http://www.constitution.ru/en/10003000-02.htm) Article 5 of Russian Constitution.

[15] YAKUTIA: AUTONOMY AND DIAMONDS (http://www.jamestown.org/single/?no_cache=1&tx_ttnews[tt_news]=6307&tx_ttnews[backPid]=218), December 1, 1996

[16] Yakutia (Russia) (http://www.crwflags.com/fotw/flags/ru-14.html) CRW Flags

[17] Amendments to the law on appointment of heads of Russia's federal subjects (http://www.rg.ru/2004/12/15/gubernatory-dok.html), 11 December 2004.

[18] Provincial government (http://www.southafrica.info/about/government/govprov.htm), SouthAfrica.info

[19] Forum of federations http://www.forumfed.org/en/federalism/by_country/index.php

[20] Federal structure of Russia (http://www.constitution.ru/en/10003000-04.htm), Article 65 of Russian Constitution.

[21] Gained independence in 1957, joined with Sabah, Sarawak, and Singapore to form Malaysia in 1963.

[22] The USSR was a federation according to the letter of its constitution, but, at least until its final years in the late eighties and early nineties of the 20th century, its governance was highly centralised in practice. See: Soviet Union section.

[23] The Socialist Federal Republic of Yugoslavia was officially proclaimed in 1963. Prior to this, the communist Yugoslav state was named Democratic Federal Yugoslavia in 1943 and then Federal People's Republic of Yugoslavia in 1946. See: Socialist Federal Republic of Yugoslavia.

Fairfax_County_Public_Schools

Fairfax County Public Schools	
Address	
8115 Gatehouse Road Falls Church, Virginia 22042	
Information	
School type	Public, school division
Founded	1870
School board	Jane K. Strauss, *Chairman* Ilryong Moon, *Vice Chairman* Elizabeth T. Bradsher Brad Center Stuart D. Gibson Martina A. Hone Sandra S. Evans Judith (Tessie) Wilson James L. Raney Kathy L. Smith Daniel G. Storck Eugene J. Coleman III, *student representative*
Superintendent	Jack D. Dale
Staff	22,779 [1]
Grades	Pre-K–12
Enrollment	177,629 [1] (2010-2011)
Campus	Suburban
Area	Fairfax County, Virginia
Website	http://www.fcps.edu/

Fairfax High School

The **Fairfax County Public Schools** system (abbreviated **FCPS**) is a branch of the Fairfax County government which administers public schools in Fairfax County and the City of Fairfax. FCPS' headquarters is located at 8115 Gatehouse Road in an unincorporated section of the county near the city of Falls Church; the headquarters has a Falls Church address but is not within the city limits.[2]

With over 175,000 students enrolled, FCPS is the largest public school system in the Baltimore-Washington and Northern Virginia Metropolitan Area. The superintendent of schools is Jack D. Dale. The school division is the 11th largest school system in the nation[1] and maintains the largest school bus fleet of any school system in the United States.

The most notable commitment the county makes to its school system is the allocation of 53.5% of its 2011 fiscal budget to the school system.[3] Including state and federal government contributions, along with citizen and corporate contributions, this brings the 2010 fiscal budget for the school system to $2.2 billion.[1] The school system has estimated that, based on the 2010 fiscal budget, the county will invest $12,820 in each student during the 2010-2011 school year.[1]

History

The public school system in Fairfax County was created after the Civil War with the adoption by Virginia of a Reconstruction-era state constitution in 1870, which for the first time in Virginia guaranteed free public education. However, during the Jim Crow era, a racially-separate system was mandated by Virginia funding laws.

Fairfax County refused to let black students attend and bused them out of the county to Manassas.[4] Despite the 1954 Supreme court ruling to end racial segregation Fairfax County Schools did not allow any black students into designated white schools until 1960.[5]

The county school system was previously headquartered at 10700 Page Avenue in an unincorporated area of the county completely surrounded by the City of Fairfax.[6] [7]

No contact rule

Kilmer Middle School, a school in the FCPS system, had a strict rule/policy of "no physical contact", meaning that contact such as high fives or hugs between friends were not allowed. The FCPS school system stood behind the rule and refused to rescind the rule. The issue was brought to light after a 13-year-old student was reprimanded for putting his arm around his girlfriend during a break, and his parents wrote to the Fairfax County School Board.[8] [9] [10] [11] [12] [13]

Debate over grading policy

Fairfax County Public Schools is known for their use of a 6-point grading scale, under which a score of 94-100 receives an A, 90-93 is a B+, 84-89 is a B, and so on. This policy has raised protests by some parents and students who support a switch to a traditional 10-point grading scale (under which 90-100 would be considered an A). FAIRGRADE, an organization started which aims to work with the FCPS research team to focus on issues that may occur because of FCPS's current grading scale, believes some areas in which students may be affected disadvantageously include:

- College admissions
- Receipt of valuable merit-based academic scholarships
- Admissions to college honors programs
- NCAA athletic eligibility
- Car insurance "good student" discounts

An example used in one report to illustrate a disadvantage of the 6 point scale noted that students in other counties can maintain higher GPA's than Fairfax County students even if they are not doing as well numerically. For example, if a student in an honors course in another county maintained a 92% average in a course for the year, he or she would receive an A and have a 5.0 to be averaged into his or her GPA, however, a Fairfax County student who maintained a 92% would get a B+ and would have a 3.5 averaged into their GPA. Saying that the significantly lower grade point value averaged into the Fairfax County student's GPA could make the student a less competitive candidate for colleges is incorrect. All of the top colleges are aware of the grading scale and are sent the appropriate documentation. So saying that it hurts a student's chances is wrong. All colleges know the difference in scales and are intelligent enough to take this into account when looking at applications. However, some college officials around the United States who have weighed in on the debate said that it would not hurt FCPS's reputation as an excellent public school system if they changed their grading policy. UCLA is among these colleges.[14]

On January 2, 2009, Superintendent Jack D. Dale announced his decision on the issue,[15] recommending the changing the weights of advanced courses but maintaining the six-point grading scale. Dale stated there was no conclusive evidence the six-point grading scale is disadvantageous for the students of FCPS. FAIRGRADE responded that "*there is still opportunity with the School Board*". -->

On January 22, 2009, the FCPS School Board directed Superintendent Dale to report back to it with a new version of the grading scale by March 2009.[16] The Board also approved changing the weighting for Honors to 0.5 effective with the 2009-2010 school year and for AP and IB courses to 1.0 retroactively.[16]

Beginning September 2009, the grading system was changed to a 10 point grading system with plusses and minuses. In this new grading system students must have a 92 for an A, 90 for an A-, 88 for a B+, and so on. Students receive a 4.0 for A's, 3.7 for A-'s, 3.3 for B+'s, and so on. Students still must reach 64 to pass.

Controversy over disciplinary policies

Fairfax County Public Schools and Superintendent Dale have been criticized for the Fairfax County school system disciplinary process, which has been blamed for the recent suicides of two students: Josh Anderson, of South Lakes High School, took his life in March 2009, and Nick Stuban, a "model student" at W.T. Woodson High School, committed suicide on January 20, 2011.[17] Stuban's family said that his spirit had been crushed in his experience with the disciplinary process.[18] In February 2011, responding to the criticism by County Supervisors Catherine Hudgins and Penelope Gross, Dale maintained that the disciplinary policy did not constitute "zero tolerance" and said that Hudgins' statements associating the disciplinary process with the suicides "for the purpose of furthering a falsehood" was "unconscionable and a blow to those who have already suffered great pain."[18]

Schools

FCPS High Schools' Rankings on *Newsweek*'s *America's Top Public High Schools*.[19] [20]

School Name	2003	2005	2006	2007	2008	2009	2010
Annandale	623	516	862	924	960	1004	1279
Centreville	177	149	99	112	106	131	141
Chantilly	72	94	162	133	126	136	224
Edison	425	692	657	694	751	798	804
Fairfax	92	196	211	157	258	133	182
Falls Church	128	289	395	1323	390	198	486
Hayfield	N/A	N/A	356	783	806	510	567
Herndon	115	174	139	161	145	166	153
Langley	31	50	81	72	55	102	118
Lee	787	1019	1215	1323	801	812	1218
Madison	787	1019	123	141	152	199	144
Marshall	184	187	201	211	254	159	203
McLean	152	128	86	76	99	115	102
Mount Vernon	305	729	636	1323	1398	1217	1516
Oakton	143	103	105	103	105	157	158

South County	N/A	N/A	N/A	N/A	143	176	308
South Lakes	407	383	608	842	515	570	221
Stuart	230	371	368	458	330	285	446
West Potomac	108	403	295	198	221	270	550
West Springfield	145	141	155	172	250	263	280
Westfield	48	126	142	179	148	192	353
Woodson	23	34	91	65	74	72	92

The *Newsweek* rankings do not include Thomas Jefferson High School for Science and Technology because it is a selective school. *Newsweek* recognized it as one of America's "public elite" high schools, having few or no average students.[21] [22]

High schools

- Annandale High School
- Centreville High School
- Chantilly High School
- Thomas A. Edison High School
- Fairfax High School
- Falls Church High School
- Herndon High School
- Langley High School
- Robert E. Lee High School
- James Madison High School
- George C. Marshall High School
- McLean High School
- Mount Vernon High School
- Oakton High School
- South Lakes High School
- J. E. B. Stuart High School
- Thomas Jefferson High School for Science and Technology
- West Potomac High School
- West Springfield High School
- Westfield High School
- Wilbert Tucker Woodson High School

Secondary schools

- Hayfield Secondary School
- Lake Braddock Secondary School
- Robinson Secondary School
- South County Secondary School

Alternative high schools

- Bryant Alternative High School
- Mountain View Alternative High School
- Pimmit Hills Alternative High School (closed)

- Woodson Adult High School

Middle schools

- Rachel Carson Middle School
- James Fenimore Cooper Middle School
- Franklin Middle School
- Robert Frost Middle School (Fairfax County, Virginia)
- Ellen Glasgow Middle School
- Herndon Middle School
- Oliver Wendell Holmes Middle School
- Langston Hughes Middle School
- Washington Irving Middle School
- Luther Jackson Middle School
- Francis Scott Key Middle School
- Joyce Kilmer Middle School
- Sidney Lanier Middle School
- Liberty Middle School
- Henry Wadsworth Longfellow Middle School
- Edgar Allan Poe Middle School
- Rocky Run Middle School
- Carl Sandburg Middle School
- Ormond Stone Middle School
- Henry David Thoreau Middle School
- Mark Twain Middle School
- Walt Whitman Middle School

Elementary schools

- Buzz Aldrin Elementary School
- Annandale Terrace Elementary School
- Louise Archer Elementary School
- Armstrong Elementary School
- Bailey's Elementary School for the Arts and Sciences
- Beech Tree Elementary School
- Belle View Elementary School
- Belvedere Elementary School
- Bonnie Brae Elementary School
- Braddock Elementary School
- Bren Mar Park Elementary School
- Brookfield Elementary School
- Bucknell Elementary School
- Bull Run Elementary School
- Bush Hill Elementary School
- Camelot Elementary School
- Cameron Elementary School
- Canterbury Woods Elementary School
- Cardinal Forest Elementary School
- Centre Ridge Elementary School

- Centreville Elementary School
- Cherry Run Elementary School
- Chesterbrook Elementary School
- Churchill Road Elementary School
- Clearview Elementary School
- Clermont Elementary School
- Clifton Elementary School
- Columbia Elementary School
- Colvin Run Elementary School
- Crestwood Elementary School
- A. Scott Crossfield Elementary School
- Cub Run Elementary School
- Cunningham Park Elementary School
- Daniels Run Elementary School
- Deer Park Elementary School
- Dogwood Elementary School
- Dranesville Elementary School
- Eagle View Elementary School
- Fairfax Villa Elementary School
- Fairhill Elementary School
- Fairview Elementary School
- Flint Hill Elementary School
- Floris Elementary School
- Forest Edge Elementary School
- Forestdale Elementary School
- Forestville Elementary School
- Fort Belvoir Elementary School
- Fort Hunt Elementary School
- Fox Mill Elementary School
- Franconia Elementary School
- Franklin Sherman Elementary School
- Freedom Hill Elementary School
- Garfield Elementary School
- Glen Forest Elementary School
- Graham Road Elementary School
- Great Falls Elementary School
- Greenbriar East Elementary School
- Greenbriar West Elementary School
- Groveton Elementary School
- Gunston Elementary School
- William Halley Elementary School
- Haycock Elementary School
- Hayfield Elementary School
- Herndon Elementary School
- Hollin Meadows Elementary School
- Hunt Valley Elementary School
- Hunters Woods Elementary School for the Arts and Sciences

- Hutchison Elementary School
- Hybla Valley Elementary School
- Island Creek Elementary School
- Keene Mill Elementary School
- Kent Gardens Elementary School
- Kings Glen Elementary School
- Kings Park Elementary School
- Lake Anne Elementary School
- Anthony T. Lane Elementary School
- Laurel Ridge Elementary School
- Lees Corner Elementary School
- Lemon Road Elementary School
- Little Run Elementary School
- London Towne Elementary School
- Lorton Station Elementary School
- Lynbrook Elementary School
- Mantua Elementary School
- Marshall Road Elementary School
- McNair Elementary School
- Mosby Woods Elementary School
- Mount Eagle Elementary School
- Mount Vernon Woods Elementary School
- Navy Elementary School
- Newington Forest Elementary School
- North Springfield Elementary School
- Oak Hill Elementary School
- Oak View Elementary School
- Oakton Elementary School
- Olde Creek Elementary School
- Orange Hunt Elementary School
- Parklawn Elementary School
- Pine Spring Elementary School
- Poplar Tree Elementary School
- Colin L. Powell Elementary School
- Providence Elementary School
- Ravensworth Elementary School
- Riverside Elementary School
- Rolling Valley Elementary School
- Rose Hill Elementary School
- Sangster Elementary School
- Saratoga Elementary School
- Shrevewood Elementary School
- Silverbrook Elementary School
- Sleepy Hollow Elementary School
- Spring Hill Elementary School
- Springfield Estates Elementary School
- Stenwood Elementary School

- Stratford Landing Elementary School
- Sunrise Valley Elementary School
- Terra Centre Elementary School
- Terraset Elementary School
- Timber Lane Elementary School
- Union Mill Elementary School
- Vienna Elementary School
- Virginia Hills Elementary School
- Virginia Run Elementary School
- Wakefield Forest Elementary School
- Waples Mill Elementary School
- Washington Mill Elementary School
- Waynewood Elementary School
- West Springfield Elementary School
- Westbriar Elementary School
- Westgate Elementary School
- Westlawn Elementary School
- Weyanoke Elementary School
- White Oaks Elementary School
- Willow Springs Elementary School
- Wolftrap Elementary School
- Woodburn Elementary School for the Fine and Communicative Arts
- Woodlawn Elementary School
- Woodley Hills Elementary School

Special education centers

- Burke Center
- Camelot Center
- Cedar Lane School
- Davis Career Center
- Key Center
- Kilmer Center
- Mount Vernon Center
- Pulley Career Center
- Quander Road School

Interagency Alternative Schools

- Adult Detention Center
- Boys Probation House
- Care Connection for Children
- Compass Day School
- Computer Enhanced Instruction Program
- Crossroads Residential Program
- Devonshire Center
- Elizabeth Blackwell School
- Enterprise School
- Falls Bridge School

- Girls Probation House
- GRANTS (GED Readiness and New Technology Skills)
- Gunston School at South County Juvenile Court
- Hickory Grove School
- Hillwood School at East County Juvenile Court
- Horizons Day School
- Independent Study Program
- Juvenile Detention Center
- Less Secure Shelter Home
- Northern Virginia Mental Health Institute
- Northern Virginia Sickle Cell Center at Inova
- Northwest Center
- Sager School
- Sojourn House
- Sunrise House
- Transition Support Resource Center
- Vista Day School at South County Juvenile Court

Transportation

FCPS operates a fleet of close to 1,600 school buses.[1] The fleet make-up consists of buses that are from 1988 to 2009. FCPS operates the following bus models:

- 1989 Blue Bird TC/2000 FE
- 1990 Blue Bird TC/2000 FE
- 1991 Blue Bird TC/2000 FE
- 1992 Blue Bird TC/2000 FE
- 1993 AmTran Genesis
- 1994 Blue Bird TC/2000 FE
- 1994 Thomas Vista
- 1994 Thomas Saf-T-Liner ER
- 1995 Thomas Vista
- 1995 Thomas Saf-T-Liner ER
- 1996 AmTran Genesis
- 1996 AmTran RE
- 1997 AmTran Genesis
- 1997 Thomas Saf-T-Liner MVP ER
- 1998 AmTran Genesis
- 1998 Thomas Saf-T-Liner MVP EF
- 1998 Thomas Saf-T-Liner MVP ER
- 1999 AmTran Genesis
- 1999 AmTran RE
- 2000 AmTran FE
- 2000 AmTran RE
- 2000 Thomas Saf-T-Liner MVP EF
- 2001 International RE
- 2001 Thomas Saf-T-Liner MVP EF
- 2002 IC RE
- 2002 IC FE

- 2003 IC RE
- 2003 IC FE
- 2004 IC RE
- 2004 IC FE
- 2006 IC RE
- 2006 IC FE
- 2007 IC RE
- 2007 IC FE
- 2008 IC RE
- 2008 IC FE
- 2008 IC CE
- 2009 IC CE
- 2009 IC RE

The transportation department is divided into 4 sections. Regular education is divided into Area 1, Area 2, and Area 3. The 4th section is Special Education. Area 1 serves the Southern part of the county, Area 2 the middle section, and Area 3 the Northern part. Special Education services the entire county.

There are three main garages that serve FCPS school buses. For service and repair, each bus is assigned to one of three garages; West Ox, Alban, or Newington.

See also

- List of Fairfax County Public Schools middle schools
- List of school divisions in Virginia

References

[1] "About FCPS" (http://www.fcps.edu/about/index.shtml). Fairfax County Public Schools. October 17, 2011. . Retrieved October 17, 2011.

[2] Home page (http://www.fcps.edu/index.shtml). *Fairfax County Public Schools.* Retrieved on April 3, 2009.

[3] (http://www.fairfaxcounty.gov/dmb/fy2011/adopted/budget_facts.htm)

[4] Russell-Porte, Evelyn Darnell (2000). "A history of education for black students in fairfax county prior to 1954" (http://scholar.lib.vt.edu/theses/available/etd-11182000-07450026/unrestricted/Final.pdf) (PDF). Doctoral dissertation. . Retrieved July 21, 2006.

[5] "Education Empire" (http://www.sunypress.edu/pdf/61145.pdf) (PDF). Albany: State university of new York Press. 2005. . Retrieved June 21, 2006.

[6] Home page (http://web.archive.org/web/20010224142554/http://www.fcps.edu/). *Fairfax County Public Schools.* February 24, 2001. Retrieved on April 3, 2009.

[7] " Fairfax city, Virginia (http://factfinder.census.gov/servlet/MapItDrawServlet?geo_id=05000US51600&_bucket_id=50&tree_id=420&context=saff&_lang=en&_sse=on)." *United States Census Bureau.* Retrieved on April 3, 2009.

[8] "Where Students Can't Hug" (http://www.time.com/time/nation/article/0,8599,1683668,00.html). *Time.* November 13, 2007. . Retrieved May 2, 2010.

[9] "CNN.com" (http://transcripts.cnn.com/TRANSCRIPTS/0706/21/pzn.01.html). *CNN.* . Retrieved May 2, 2010.

[10] "Students Feel the Squeeze as Schools Ban Hugs" (http://www.foxnews.com/story/0,2933,309296,00.html). *Fox News.* November 8, 2007. .

[11] http://www.msnbc.msn.com/id/19293872/

[12] The Washington Post (http://www.washingtonpost.com/wp-dyn/content/article/2007/06/17/AR2007061701179.html)

[13] http://www.guardian.co.uk/world/2007/jun/19/usa.schools,NPR http://www.npr.org/templates/story/story.php?storyId=11181177

[14] "Does Fairfax Give Fair Grades?" (http://www.connectionnewspapers.com/article.asp?article=318883&paper=73&cat=104). . Retrieved 12-6-2008.

[15] Dale, Jack (January 8, 2009). "Review of Grading Regulations" (http://www.boarddocs.com/vsba/fairfax/Board.nsf/d62d9cb847ef1cbd87257328006795e4/39ceb62d5a0b1726872574d3005d63d7?OpenDocument). *Fairfax County Public Schools School Board eGovernance System.* . Retrieved January 24, 2009.

[16] Sabo, Linda (January 23, 2009). "School Board Action 1-22-09" (http://www.boarddocs.com/vsba/fairfax/Board.nsf/News/2147F625F9041EAE8525754700172769?OpenDocument). *Fairfax County Public Schools School Board eGovernance System.* . Retrieved

January 24, 2009.

[17] "Suicide turns attention to Fairfax discipline procedures" (http://www.washingtonpost.com/wp-dyn/content/article/2011/02/19/AR2011021904528.html). The Washington Post. February 20, 2011. . Retrieved February 22, 2011.

[18] "School Superintendent Jack D. Dale defended Fairfax County's discipline policies" (http://www.washingtonpost.com/wp-dyn/content/article/2011/02/11/AR2011021106160.html). The Washington Post. February 11, 2011. . Retrieved February 22, 2011.

[19] Newsweek's *America's Top Public High Schools* for 2009 (http://www.newsweek.com/id/201160)

[20] Mathews, Jay (May 18, 2008). "FAQ: Best High Schools" (http://www.newsweek.com/id/137415). Newsweek. . Retrieved September 9, 2008.

As noted in the FAQ, the ranking is based on Jay Mathews' experience, does not account for how well students do on the test or thereafter.

[21] Pace, Gina (May 18, 2008). "Best High Schools: The Public Elites" (http://www.newsweek.com/id/137421). Newsweek. . Retrieved November 30, 2009.

[22] Matthews, Jay (June 9, 2009). "Top High Schools: The Public Elites" (http://www.newsweek.com/id/201138). Newsweek. . Retrieved November 30, 2009.

External links

- Fairfax County Public Schools (http://www.fcps.edu/)
- Archives of fcps.k12.va.us (http://web.archive.org/web/*/http://www.fcps.k12.va.us)

Article Sources and Contributors

Nitin_Pradhan *Source*: http://en.wikipedia.org/w/index.php?title=Nitin_Pradhan *Contributors*: 2MuchCoffee, Armbrust, Beachcruizer, Bgwhite, DaffyBridge, Giraffedata, Gov11234, Hmains, Malcolma, Ponydepression, Rich Farmbrough, TheDatel, WOSlinker, WereSpielChequers, 9 anonymous edits

Chief_information_officer *Source*: http://en.wikipedia.org/w/index.php?title=Chief_information_officer *Contributors*: 72Dino, Aecis, Aldie, Alphachimp, Andrewmin, Aranea Mortem, Benjamin Mako Hill, BlueSkyNoRain, Bruce404, Buster7, C172rg, CIOGuru, Cebess, ChristianBk, Comfychaos, Cybercobra, Dandv, Daniel73480, David Shay, Davidp, Dismas, Diza, Djharrity, Dmccreary, Dub617, Dukeofgaming, EFormsBlogger, Egil, Ehheh, Fbarw, Gadfium, Gaia Octavia Agrippa, Gatemansgc, Greyengine5, Groverxv, Hede2000, Hqb, Ian Pitchford, Iswive, Iznogoud, JForget, Jamalystic, Jaysee21, Jazmatician, Jmilestaylor, Johnloia2112, JonHarder, Joseph.strunz, Jplarkin, Julsssark, Jvlock, Kuru, Laran.evans, Leslie Mateus, LiDaobing, Lotje, Mairi, Millstream3, Mitchello, Mote, Mpeisenbr, MrOllie, Neutrality, Nslonim, Nurtsio, Ohnoitsjamie, OminousSkitter, OnceAlpha, OneGyT, Pan Dan, PaulHanson, Poweroid, Pravin.taskseveryday, Rhobite, Rich Farmbrough, Rjwilmsi, Robert Merkel, Robofish, Rock2e, Ronz, Rrastin, Schneelocke, Sebesta, Shajela, Smpclient, SusanLesch, Svetovid, Swineinfluenza, Tellmewhy1, Tempodivalse, The Thing That Should Not Be, Thecheesykid, Themfromspace, Thiagoeh, Thomasjl, Vague, Wiki3 1415, Woohookitty, Yarnalgo, Zelig123456, Zondor, ישׁ דוד, 145 anonymous edits

United_States_Department_of_Transportation *Source*: http://en.wikipedia.org/w/index.php?title=United_States_Department_of_Transportation *Contributors*: Aesopos, AgnosticPreachersKid, Ajraddatz, Amakuru, Anders.Warga, Annelise.d, Arcadian, Ardfern, Asmallwhitecube, Aude, Bhuck, Billinghurst, Bissinger, BrendelSignature, Burgundavia, CD972, CWenger, Cadby Waydell Bainbrydge, Cag1970, CapitalR, Ched Davis, Chickyfuzz123, Collins Robert Cramer, Connormah, Coolcaesar, CopperSquare, Creationlaw, Curps, Dancing is Forbidden, Danielarce, DavidLevinson, Deh, Ed g2s, Electron9, Epolk, Ervaude, Esrever, Flyguy33, GageSkidmore, Gary King, GearedBull, Gint'ire, Gobonobo, GoldRingChip, Gov11234, Grafen, Ground Zero, HiB2Bornot2B, Hoshie, IbLeo, Int21h, J.delanoy, Jaan513, Jakobdoerr, Jaqu, Jatkins, Jbenjos, Jengod, Jeronimo, Jslynch, Klugrimsky, Koavf, Koward, Kross, L Kensington, Leuliett, Levineps, Lightmouse, Luckyhalo, Mac, Malepheasant, MaryD99, Minesweeper, Modster, Motthoangwehuong, Mulad, Neutrality, Nightkey, Nopetro, Ocedits, Omicronpersei8, Patrick, Pats1, Pearle, PennySpender1983, Postdlf, Ppntori, Reywas92, RicJac, Rising*From*Ashes, Rodhullandemu, Rougher07, Rrenner, SQL, Safetyinspector, Sardanaphalus, Scbalik, Smith03, Socby19, SomeUsr, Sross (Public Policy), Tabletop, Template namespace initialisation script, The Rambling Man, Thewinchester, Tinton5, Tobias Hoevekamp, TommyBoy, UpstateNYer, Utmostevil, Vegaswikian, Vrenator, WhisperToMe, Wilsonchas, Xundeadpkerx, Yuriybrisk, Zereshk, 134 anonymous edits

Presidency_of_Barack_Obama *Source*: http://en.wikipedia.org/w/index.php?title=Presidency_of_Barack_Obama *Contributors*: 101090ABC, 3328lzc, 4wajzkd02, 6tg8, 712emgz, 781fcg, 871lacc, A waitress in a cocktail bar, A. Parrot, A16529, A6702, Aaron charles, Abrazame, Acps110, Adri K., Afranelli, AgnosticPreachersKid, Alan Liefting, Alansohn, Albertod4, AlexanderHaas, Ali'i, Allknowingjipsey, Allstar86, Allstarecho, Amalthea, Ammon86, Andrew C. Bowman, Andrewlp1991, Andy Marchbanks, Andycjp, Anirvan, Anon1006, Antandrus, Arman Cagle, Armyrifle9, As clumsy as I am stupid, Ashershow1, AuroraborealisCobra, Austin512, Avala, B Milnes, BD2412, BTsm4bKG, BalticPat22, Banaticus, BarryTheUnicorn, Baseball Bugs, Bassbonerocks, Belle jolie, Ben2us, BenZern, Benny45boy, Bewebste, Bigchickenhead, Bigtimepeace, Bill0919, BillC, Biruitorul, Bixby becomes Ferrigno, Bjkijkjr83, Blackchaosmage19, Blanchardb, Bleemsz, Blert, Blubberboy92, Bob bobato, Bobblehead, Bobby 538, BomBom, Boncardoaward, Bongwarrior, Bonsrox, Boromir123, Bou67, Brothejr, Brozyp, Bruce1ee, Bry456, Bsadowski1, Bsimmons666, Bubbavspigman, Bullfroglover09, CZmarlin, CaelumArisen, Calabe1992, Calmer Waters, Calorus, Captain Lance Murdoch, CardinalDan, Carolmooredc, Cassandro, Cdanek, Cdg921, Centralkong, Cfortunato, Cgersten, Chardish, Chaser (away), Chess King Winner, ChildofMidnight, Chris the speller, Chrishmt0423, Chuck Marean, Ciphergoth, Clap2304670327509345, Clsin, Cmguy777, Coemgenus, CompProjects, Conman33, Connelly, ConstantineChernabog, Cosmic Latte, CouldOughta, Cr7aJ, Crd721, Crunch, Cube lurker, D19zo, D99figge, DD2K, DIEXEL, DJ Clayworth, DLCruise, DMacks, DRosenbach, Dabomb87, Dabosnian, Dallasmartino, Dard Lord XXXIII, Darth Kalwejt, Davehi1, Dayewalker, Deadwind20, Debresser, December 3, Dechm4af, Deckiller, Decora, Defaulter.passi, Demonic224, Denisarona, Deytel, Diaa abdelmoneim, Diddlyman2004, Die4Dixie, DigbyDalton, Dimevader88, Dionyseus, Dlohcierekim, Douglemeister, Dour2, Dr.enh, Dragged up from the underworld, Drilnoth, Drlight11, DubaiTerminator, Dudemanfellabra, Durga Dido, Dwainwr123, EKMichigan, EagerToddler39, Edinburgh Wanderer, Edokter, Edward130603, Edwardlalone, EhJJ, Eight 67 5309, Eisfbnore, Elliotteque, Eloquence, ElzbthMllr, Embryomystic, Enginear, Epeefleche, Epson291, EricMiles, Ess5309, Excirial, FT2, Falcon8765, Fat&Happy, Favonian, Fetchcomms, Fh7aTP8F, Fieldday-sunday, Fifty7, Fightin' Phillie, Firsfron, Floydian, Forgot the Password, FrankieG123, Fratrep, Free Bear, Frietjes, Frotz, Furtive admirer, Fzwyj72, Fæ, G-MMDC, GRB080, GageSkidmore, Gagz135, Gary King, GcSwRhIc, Gogo Dodo, GoingBatty, GoneAwayNowAndRetired, Gordon Ecker, GorillaWarfare, Grant.Alpaugh, Greatwords662, Ground Zero, Grsz11, Grundle2600, Guettarda, Gurch, Guyzero, Gyrobo, Gz33, Hadrianheugh, Haha169, Hand me the keys you fairy godmother!, Harro5, Headbomb, Hekerui, Hello snotface!, Herculoids, Hgxzm23, Hh224, HiLo48, Hibbertson, Hipocrite, Hl, Hmains, Honkeysdonkeysmonkeys, Hoshie, I came out of lightspeed too close, I think I'll dye my hair blue, I'm gonna pack my bags and FLY, I9c5t3, Iamwisesun, Ian Burnet, Igoldste, Imagination's got the best of you, ImperfectlyInformed, InfoFan, InterwebUsr, Iota, Iridescent, Isfisk, Ism schism, J.delanoy, J04n, JDawg105, JEN9841, JForget, JOHNDOEZ123, JTSchreiber, JZX 527, JackofDiamonds1, JahnTeller07, JamesMLane, Jao, Jatkins, JayJasper, Jayron32, Jc-S0CO, Jdweller, Jessedow24, Jivee Blau, Jmundo, John of Reading, Johnny Spasm, Jojhutton, Joker123192, Jomizell, Joseph Solis in Australia, Joshdboz, Jusdafax, JustGettingItRight, Justmeherenow, KRBM, KSWarrior8, Ka Faraq Gatri, Kakofonous, Kaldari, Karenjc, Kelapstick, Kevinp2, Khoikhoi, KillerChihuahua, Kinaro, Kingpin13, Kirsted, Klilidiplomus, Korath, Kudzu1, Kungfuactionjesus, Kuralyov, LCahill, Laaknor, Lagelspeil, Lampman, Larcen357, Lerchdude, LesHurv, Levineps, Lincolnite, Lionelt, Lithoderm, Little runt sitting there on a log, Lngt1124, Lobosland, Lols107, Loodog, Loonymonkey, LtNOWIS, Lumaga, Luna Santin, Lwalt, MONGO, MOOSENUT42, Machetemirage, Madkayaker, Madlee, Magnum! This is all your fault!, Major Joshua Sleestak, MakeBelieveMonster, Marek69, Marmandy, Marshall Williams2, Martin Raybourne, Martin451, Materialscientist, MattGiuca, Mayur, McSly, Mcguemcgueyouyou, Medo4, Melissa 3383, Mike-Kerkhoven, Mild Bill Hiccup, Millionsandbillions, Misbeliever, Mlpearc, Mnnlaxer, Moneyshea, Monty845, Movieguruman, Mr. Anon515, Mr. Jackson 1952, Msclguru, MuZemike, Muboshgu, N5iln, NYyankees51, Nasnema, NawlinWiki, NewDutchy, NewEnglandYankee, Newross, Niceguyedc, Nightscream, NonvocalScream, North8000, NortyNort, Notchbag, Novem Linguae, NuclearWarfare, ONEder Boy, Obamiac, Odie5533, Ohconfucius, Ohms law, Olegwiki, Onze11, Orange Suede Sofa, Originalwana, Ottre, OuroborosCobra, Outside Woman Blues, P. S. Burton, PRRfan, Parmadil, Paulbata80, PbQAms, Pdanbom95, Peace and images of pluto, Peas 447, PericlesofAthens, Peter Karlsen, Peterhansen2032, Pexise, PhGustaf, Philip Trueman, Phoenix and Winslow, Phynicen, Physalia physalis, Piano non troppo, Pitt, Plau, Plumber, Plumkin001, Politicalysmart, Politics2012, Preivet, Primal Wind, Princess Caraboo, Prolog, Psantora, Put Captain Solo in the cargo hold, Qetuo1470, QueenCake, QueenofBattle, Qwyrxian, Qxsp9h, R6axa, RA0808, Rabbit Seasoning, RandomHero129, Raprchju, Ratemonth, RayneVanDunem, Rd232, Reachblue2245, Rebroad, Regancy42, Reywas92, Rich Farmbrough, Rickyrab, Rjking91, Rjwilmsi, Rmosler2100, Rogerd, Ronhjones, Rontrigger, Rooster hits the washboard, Rostz, Rougher07, Rrburke, Rrius, Rt64133, Runningonbrains, Rural9625drink, S51438, SChaos1701, SGGH, SMP0328., Saddhiyama, Salamurai, Sandstein, SarekOfVulcan, Scapler, Sceptre, Scjessey, Scoops, ScottSteiner, Scottst, ScottyBerg, Sean Quixote, Sean.hoyland, Seaphoto, Sf46, Shadowjams, Shazisg123, Shinderj, Shirik, Showtime2009, Sight1234, SimonATL, Sirat2, Sjones23, Skalman, Skiing in Switzerland, Skizzik, Skomorokh, Sln3412, Sloane, Sm8900, Smk436, Snowmanb, Solopiel, Spaghett89, Spandlingford, Speedster239, Spellage, Spinach Monster, Spitfire, Splette, Stephenjordanisawesome, Sterster101, SteveSims, Steven J. Anderson, Steveng72, Stickguy, Stillwaterising, Stonewallhenry, Strait, Student7, Stupidguy24, SummerPhD, Sumone10154, SuperDaveMusic, SuperMoonMan, SusanLesch, SwisterTwister, Swng1122, THENAMESGGBABY, TJRC, Tabletop, TaerkastUA, Tarc, Tatoie, Tbhotch, TeneeshaJackson, Terra Novus, Terrierhere, Tgeairn, Th1rt3en, Thatguyflint, The Cunctator, The Fwanksta, The Magnificent Clean-keeper, The Pony puts us in a trance, The Red Peacock, The Sartorialist, The Squicks, The Thing That Should Not Be, The days are dull the nights are long, The news said it's raining in New York, The sock that should not be, TheCodeman4, TheThomas, Themfromspace, Theshibboleth, Thingg, Thoroughgoodness, Thoughts and dreams that scatter, Threeafterthree, Tide rolls, Tigerghost, Timneu22, Timtrent, Tinmanic, Tktru, Tobby72, TomCat4680, Tommy2010, Tony Fox, Tony Sidaway, TonyTheTiger, Tpbradbury, Traxs7, Treybien, Tricks Are Not Treats, Trumpetrep, TwisstedVisionz, Ulric1313, Umbralcorax, Unitanode, UnqstnableTruth, Untilzero, VBGFscJUn3, Valador, Valerieb09, Vanished user 03, Verizonwireless666, Vicenarian, Victorgrigas, Vladko, VolatileChemical, Vrenator, Vrysxy, Vulgarian Visigoth, Wai Hong, Wasn't Tomorrow Wonderful?, Wasted Time R, West.andrew.g, Wforte, Whywhenwhohow, Wiki13, WikiPuppies, Wikidemon, Wikiman699, Wikipeterproject, Wikiposter0123, William Jockusch, William S. Saturn, Wine Guy, Winterus, Wknight94, Woogee, Wtmitchell, X!, XCentristFiasco, Xavexgoem, Xhaoz, YUL89YYZ, Yachtsman1, Yellowdesk, YeshuaDavid, Yodalovesyou, Yoshoku, You sunk my battleship!, You used to call me paranoid, Yumad 2011, Zaarah10, Zachlipton, Zenswashbuckler, Zrbtt, Zzuuzz, 742 anonymous edits

United_States_Secretary_of_Transportation *Source*: http://en.wikipedia.org/w/index.php?title=United_States_Secretary_of_Transportation *Contributors*: Adam sk, Ando228, Bad Graphics Ghost, BarretB, Bonadea, Borgx, Caper13, CapitalR, Celestra, Chris the speller, Chrishmt0423, Chrisn4255, CommonsDelinker, DMG413, Delldot, Diberri, Dkanem, Docu, Dough4872, Eddy23, Epolk, Fdewaele, GageSkidmore, Golgofrinchian, GoodDay, Hirokun, Hoary, Hokeman, Hoshie, JPMcGrath, Jack Cox, Jjmillerhistorian, John K, Katagelophobia, Kbdank71, Klugrimsky, Kudzu1, Levineps, Luk, MartinHarper, Michelet, Minesweeper, Minutiaman, MrDolomite, Nate Silva, Nopetro, NuclearWarfare, Nutiketaiel, Pearle, Postdlf, Qxz, Reywas92, Rrius, Ruslik0, SarahStierch, Scohoust, Scott5114, Shoessss, Simbainkempner, Snappy, Stormie, TCY, Tedickey, Tom, TommyBoy, TwizteDope, VerruckteDan, Vrysxy, WaitingForConnection, Weirdoactor, William Saturn, 76 anonymous edits

Ray_LaHood *Source*: http://en.wikipedia.org/w/index.php?title=Ray_LaHood *Contributors*: 101090ABC, 5 albert square, Abrazame, Adam Carr, Alansohn, All Hallow's Wraith, Americus55, Anders.Warga, Anomalocaris, Autumn Hawk, BD2412, Bahua, Barakishidan, Beland, Ben76266, Bender235, Benore, Blubberboy92, Bluedogtn, Brendan Moody, Briaboru, BrokenSphere, Byrdin2006, CB2379, CRKingston, Captain-tucker, Celestra, Cgingold, Chaser, Closeapple, CommonsDelinker, Cyril Washbrook, D6, Dallas Hays, Dav301, Daveman 84, Dblevins2, Dhartung, Disneyfreak830, Downtown dan seattle, Dr who1975, Drilnoth, Duffy2032, EATC, Eastlaw, Edmonton7838, EmmaRoad, Epbr123, Everyking, Exchaoordo, Fdewaele, Flatterworld, Flugzeug fünf, Foofighter20x, GageSkidmore, Gang14, George Al-Shami, Gidonb, Gilliam, Goethean, GoneAwayNowAndRetired, GoodDay, Griffinofwales, Ground, Gunsnroads, HUZZAH HANUKKAH, Has164, Hekerui, Hmains, Hu12, Iamwisesun, Im fat08, JDP90, JForget, Jack Cox, JamesAM, JamesB3, JamesMLane, JayJasper, Jdhowens90, Jengod, Jesuschex, Jim.henderson, Jmchuff, John of Reading, Jonathan.s.kt, Jonathan321, Jrcla2, Jun Nijo, Justmeherenow, KConWiki, Kbdank71, Kbir1, Kenallen, KeptSouth, Kudzu1, Kuralyov, Kyleandrew1, Levineps, Lightmouse, Louturks, Lyoshka, Mack2, Meelar, Minimac, Motorhead429, Mrh30, NatusRoma, Noformation, Normjaso, Oreo Priest, PhilGutis, Politicsislife, PollShark, Postdlf, Propol, Ptolemy Caesarion, Rankun, Reaper Eternal, Reywas92, Rich Farmbrough, Rockyobody, Romeisburning, Rontrigger, Rpogge, Rrius, Rslnerd, Rushadthomas, Ryou-kun16, ScooterDe, Seth Ilys, Seyon, Sfgiants906, Sholom, SnowFire, Snowolf, Sperril, Stemonitis, Switzpaw, Sylong, Tabletop, TastyPoutine, Tdl1060, TheHonMP, Therefore, Therequiembellishere, Thismightbezach, Tktru, Tom, Tommy Nero Sullivan, TommyBoy, TonyTheTiger, Treybien, Tvoz, Ulric1313, UpstateNYer, Veritémaster, Vic Troy, VitaleBaby, Vondicus, Voot42, Vystrix Nexoth, Willcoop, Xstryker,

Yuber, Zenohockey, Zjhafeez, Zpalmese, Zsero, شررز, 172 anonymous edits

Federation *Source*: http://en.wikipedia.org/w/index.php?title=Federation *Contributors*: (, 28421u2232nfenfcenc, 52 Pickup, ATG, Aaron of Mpls, Abshirdheere, Adam Carr, Ahen1020, Aiman abmajid, Ajlmoraes, Alansohn, Alexius08, Algri, Amorim Parga, Anaxial, Andrew Gwilliam, Andrewfrueh, Andrzej Kmicic, Andy120, Angela, Anthares, Antonio Lopez, Appraiser, Arctic.gnome, Arsonal, Ascidian, Asdofindia, Astrofreak92, Atob, AusSF, Auxodium III, Avjoska, Avs dps, AzaToth, B6utle5r, Badanedwa, Baron von Chickenpants, Bastin, Baybruh510, Beano, Bebenko, Beland, Bhumiya, Big Adamsky, BilCat, Bjmurph, Blind Man Walking, Blue-Haired Lawyer, Bnjohn, Bobbo, Bobbra, Bobo192, Bolivian Unicyclist, Bongwarrior, Boofhead185, Bosonic dressing, Briaboru, C mon, CJ, Caeruleancentaur, Cafzal, Canationalist, Carlroller, Cassowary, Centro Studi sul Federalismo, Ch0002, Chealer, Chris 73, Chris the speller, Civil Engineer III, Closedmouth, Colipon, CommonsDelinker, Conorbrady.ie, Crazynas, CuboneKing, Curiosity1, Cynical, D. F. Schmidt, D6, DB 103245, DIGIwarez, DSuser, DanMS, Dantadd, Daveb, David91, DavidA, Dawn Bard, Dbachmann, Dbenbenn, Ddye, Dekimasu, Demicx, DerHexer, Dewritech, Dissident, Doc Fabulous, Doc Quintana, Docu, Donarreiskoffer, Doulos Christos, Dpaajones, Duffman, Dumelow, Dynamization, Dysepsion, Dúnadan, E Pluribus Anthony, Eagle4000, Earth, EdJohnston, EdwinHJ, Electionworld, Eluchil404, Ender's Shadow Snr, Enlip, Erik9, Estevoaei, EuTuga, Fabartus, Favonian, Fingerz, Fishmech, Flyers13, Footwarrior, Frankenpuppy, Fredrik, Freigut, Fremfreak23, FreshFruitsRule, FrumpyFanny77, Fudoreaper, Garrett54540, Gary King, Gawaxay, George2001hi, Gerald Farinas, Gilliganisland123, Glenn, Glorfindel Goldscheitel, Go get away forever, Goatasaur, Golbez, Grant65, Grantbonn, GreatWhiteNortherner, Gregbard, Gregorof, Gribbles, Groudon185p, Grover cleveland, Grsz11, Guðsþegn, Gvogas, Gwernol, Haham hanuka, Haipa Doragon, Hairy Dude, Hallward, Hamish Lawson, Harjk, HdZ, Hectorian, HexaChord, Hobartimus, Holger1076, Htonl, Huhsunqu, Husond, IANVS, Ianjm, Icairns, Imc, Imhotikhtokep, Instantnood, Iota, Iterator12n, Ivan Volodin, JLH, JaGa, Jagerz00, James.S, January, Jao, Jay, Jebba, Jeff3000, Jeroen, Jiangji, Jimc, Jklin, Johann Grimm, John of Reading, Joruul, Joseph Solis in Australia, Joshua, José Fontaine, Jsc83, Julesd, JustPhil, Karl-Henner, Katalaveno, Kdmitry, Keenman76, Khukri, Kintetsubuffalo, KirkEN, Kkm010, Ko-Ka, KokomoNYC, Kookyunii, Koov, Krich, Kurlso360, La goutte de pluie, Lacrimosus, Lanççelot, Le Anh-Huy, Ledmonkey, Leflyman, Lendu, LilHelpa, Limongi, Lithic, Llywrch, Lmc169, Logologist, Lokal Profil, Lordscissorhand, Loureiroandre, Lubiesque, Lucky13pjn, Luxem, Lysy, MBisanz, MacGyverMagic, Malo, Man vyi, Mani1, Marek69, Masseman, MauriManya, Mb nl, Menchi, Mentifisto, Mhdmahbubmiah, Michael Johnson, Middayexpress, MikeLynch, Mikki mouse, MileyDavidA, Miraculouschaos, Mistakefinder, Mitth'raw'nuruodo, Monsopiad, Montrealais, Mr Gronk, N2e, Natkeeran, Nazrani, Neo-Jay, Nepenthes, Nickshanks, Night w, Nightstallion, Nihil novi, Nikodemos, Nil Einne, Npeters22, NuclearWarfare, Nurg, Ojw, On the other side, Onebyone, Oo64eva, Orange Suede Sofa, Otsego, Oxymoron83, PCHS-NJROTC, PGWG, Paedia, ParticleMan, Paul Benjamin Austin, Pavel92, PaxEquilibrium, Peace keeper II, Perceval, Persian Poet Gal, PeterEasthope, PeterTrompeter, Peterl, Pgreenfinch, Phil w jones, Picapica, Pierpietro, Piotrus, Pne, Poldavo, Prodigy0987, Pyroclastic, Pádraic MacUidhir, Qst, Quiensabe, R'n'B, RA0808, RL0919, Raeky, Rajan R, Raul654, RazorICE, Rbanzai, Rcgy, Red King, Renata3, Returningvantal, Ricky@36, RobertMfromLI, Robertgreer, Robteix, Rohlib, Rolgn, RomeW, Ronhjones, Rr parker, Rror, Rupertslander, Ruppee, SHIMONSHA, Sam Blacketer, Sam Hocevar, Samwaltz, Sardanaphalus, Scetoaux, Scott Gall, SeNeKa, Seba, Sesel, Sesu Prime, Shade 5154, Shanes, Sherwoodseo, Shoreranger, Silverhorse, SimonX, Singhalawap, Siriaeve, Sivanesh, Skouratov, Sleigh, Small.is.powerful, Sobolewski, SorryGuy, Spesh531, SpicyPanda, Spinach Monster, Spinningspark, Ssolbergj, Stephen Bain, Stephenb, Supertask, Sw7891, SweetNeo85, Synchronism, TFMcQ, Technolefty, Teryx, Tgv8925, The Nerd from Earth, The Thing That Should Not Be, ThePedanticPrick, Theaznlaw, Thebeastofnanking, Themoodyblue, Thirty-seven, Thomas Paine1776, Thomaslw, Thumperward, Tiddly Tom, TigerShark, Tommy Kronkvist, Tomtom, Tomtom9041, TopoChecker, Toussaint, TreasuryTag, Treisijs, Tuyi, UKWiki, Urbanus Secundus, Useless Fodder, Uxejn, Valip, Vanka5, Velimir85, Versus22, VictorAnyakin, Void180, Vzbs34, Washdivad, Webridge, WereSpielChequers, Wertuose, White Shadows, Wild ride, Wilfried Derksen, William Avery, Wisdom89, Woohookitty, Wtmitchell, XLR8TION, Xerex, Yakoo, Yardcock, Yeu Ninje, Yggdrasil, Yvwv, ZaacB, Zachorious, ZayZayEM, Zfr, Zhonghuo, Zlerman, Zondor, Zoris Trömm, Zzuuzz, , 702 anonymous edits

Fairfax_County_Public_Schools *Source*: http://en.wikipedia.org/w/index.php?title=Fairfax_County_Public_Schools *Contributors*: 1995hoo, 1q2w12, 20176, Acntx, Acroterion, Alansohn, Alexius08, Alias Flood, AliveFreeHappy, Andypandy.UK, Ankit.pigs, Arsonal, Arvinahmadi, BLM Platinum, BaronLarf, Beigefood, Bento00, BlackbeardK, Blacksonicus, Bobo192, Boiler321, Brendanconway, Broker of darkness, Cerebus19, CheetahMan1, Connormah, Coredesat, DARTH SIDIOUS 2, DJMcNiff, DVdm, Dabja, Dale Arnett, Discospinster, Dtran, Duchess mommy, EEMIV, Eatmyfetustwo, Elesi, Erik9, Excirial, FCPS ICRE 684, FeralitYzERO, Ferbess, Fitzwrite2, FlamingSilmaril, Floquenbeam, ForgottenManC, Freddiem, Funeral, GTBacchus, Galvertez, Garzo, GoingBatty, Guruoat, Gwernol, Hailey C. Shannon, HamburgerRadio, Heynow09, Howdybob, Interchange88, Invalid edition, Ixfd64, JJay, Jaeiscool, Jaksmata, JamesBWatson, JamesHL2009, Jamesr66a, Jauerback, Jeff Faust, Jj137, Jllm06, Joelmills, Jrbarry, JustAGal, Kazak, KeepOnTruckin, Keskiyonaurinko, Kingturtle, Klmarcus, Kraftlos, Kross, LASA Scramblers, Laura9355, Lonestar662p3, Mitsukai, MrKIA11, Nburden, No1lakersfan, Noroton, Nzo9, Oxymoron83, Pauli133, Paxswill, Peripitus, Peter Chastain, Peterpeterpan, Phenominal1999, Phil Boswell, Pigs.ankit, Plothands, Polyglot-Mom, Pumpkinart, Quarl, Racepacket, Redskull619, Resurgent insurgent, Rich Farmbrough, Rjwilmsi, Robbie098, RossPatterson, Rsm99833, RyanGerbil10, Ryanaxp, SJP, SWAdair, SaveThePoint, Ser Amantio di Nicolao, Shizzlefofizzle, SpLoT, Ssj4android, SterlingNorth, Student7, Stupidkrazykarl, Sumergocognito, Superm401, Suvablee0506, Tabletop, Tcncv, Tedickey, The Universe Is Cool, TheRanger, Thepangelinanpost, Thivierr, Timsabin, Tony Sidaway, Tstahmer, VT hawkeye, Vaoverland, Vegaswikian, WAWEEZ, Watchreader, Waterfalls12, West Brom 4ever, WhisperToMe, Whylimeemily, WikiXan, Wolfboy93, Xiong Chiamiov, Xnatedawgx, 317 anonymous edits

Image Sources, Licenses and Contributors

File:DOT IT Portfolio Rationalization.jpg *Source*: http://en.wikipedia.org/w/index.php?title=File:DOT_IT_Portfolio_Rationalization.jpg *License*: unknown *Contributors*: TheDatel

File:Digital Transportation Exchange (DTE).jpg *Source*: http://en.wikipedia.org/w/index.php?title=File:Digital_Transportation_Exchange_(DTE).jpg *License*: unknown *Contributors*: TheDatel

file:US-DeptOfTransportation-Seal.svg *Source*: http://en.wikipedia.org/w/index.php?title=File:US-DeptOfTransportation-Seal.svg *License*: unknown *Contributors*: User:Sakurambo

File:Usdot headquarters.jpg *Source*: http://en.wikipedia.org/w/index.php?title=File:Usdot_headquarters.jpg *License*: unknown *Contributors*: User:Aude

File:US Dept of Transportation 1967 Official Seal.jpg *Source*: http://en.wikipedia.org/w/index.php?title=File:US_Dept_of_Transportation_1967_Official_Seal.jpg *License*: unknown *Contributors*: US Government

File:Barack Obama signature.svg *Source*: http://en.wikipedia.org/w/index.php?title=File:Barack_Obama_signature.svg *License*: unknown *Contributors*: User:SyntaxError55

File:Barack Obama Inauguration.jpg *Source*: http://en.wikipedia.org/w/index.php?title=File:Barack_Obama_Inauguration.jpg *License*: unknown *Contributors*: Steve Jurvetson from Menlo Park, USA

file:President Barack Obama with full cabinet 09-10-09.jpg *Source*: http://en.wikipedia.org/w/index.php?title=File:President_Barack_Obama_with_full_cabinet_09-10-09.jpg *License*: unknown *Contributors*: The Official White House Photostream

File:Obama and Bill Clinton.jpg *Source*: http://en.wikipedia.org/w/index.php?title=File:Obama_and_Bill_Clinton.jpg *License*: unknown *Contributors*: The White House from Washington, DC

File:20090124 WeeklyAddress.ogv *Source*: http://en.wikipedia.org/w/index.php?title=File:20090124_WeeklyAddress.ogv *License*: unknown *Contributors*: The United States Government

File:Dmitry Medvedev 1 April 2009-1.jpg *Source*: http://en.wikipedia.org/w/index.php?title=File:Dmitry_Medvedev_1_April_2009-1.jpg *License*: unknown *Contributors*: Presidential Press and Information Office

File:Barack Obama Michelle Obama Queen Elizabeth II Buckingham Palace London.jpg *Source*: http://en.wikipedia.org/w/index.php?title=File:Barack_Obama_Michelle_Obama_Queen_Elizabeth_II_Buckingham_Palace_London.jpg *License*: unknown *Contributors*: White House (Pete Souza) / Maison Blanche (Pete Souza)

File:Obama and Biden await updates on bin Laden.jpg *Source*: http://en.wikipedia.org/w/index.php?title=File:Obama_and_Biden_await_updates_on_bin_Laden.jpg *License*: unknown *Contributors*: AnonMoos, Carnby, Chaser, Chrishmt0423, Closeapple, Danim, Delaywaves, Dismas, Docu, EurekaLott, Face, Fitoschido, FlickreviewR, Goodvac, Infrogmation, Jonathan Schilling, L'amateur d'aéroplanes, Liesel, Magog the Ogre, NYC JD, NotFromUtrecht, Peter Weis, Pilettes, Postdlf, Surya Prakash.S.A., TCY, Tabercil, Túrelio, UpstateNYer, Ysangkok, 13 anonymous edits

File:President Speaks to Crew on Space Station.jpg *Source*: http://en.wikipedia.org/w/index.php?title=File:President_Speaks_to_Crew_on_Space_Station.jpg *License*: unknown *Contributors*: NASA/Bill Ingalls

File:President Obama speaks at Kennedy Space Center.jpg *Source*: http://en.wikipedia.org/w/index.php?title=File:President_Obama_speaks_at_Kennedy_Space_Center.jpg *License*: unknown *Contributors*: NASA/Jim Grossmann

File:Beer summit cheers.jpg *Source*: http://en.wikipedia.org/w/index.php?title=File:Beer_summit_cheers.jpg *License*: unknown *Contributors*: Official White House Photo by Pete Souza

Image:US-DeptOfTransportation-Seal.svg *Source*: http://en.wikipedia.org/w/index.php?title=File:US-DeptOfTransportation-Seal.svg *License*: unknown *Contributors*: User:Sakurambo

File:Ray LaHood official portrait.jpg *Source*: http://en.wikipedia.org/w/index.php?title=File:Ray_LaHood_official_portrait.jpg *License*: unknown *Contributors*: Department of Transportation

File:Alan_stephenson_boyd.jpg *Source*: http://en.wikipedia.org/w/index.php?title=File:Alan_stephenson_boyd.jpg *License*: unknown *Contributors*: DOT, Office of the Historian

File:Volpe.gif *Source*: http://en.wikipedia.org/w/index.php?title=File:Volpe.gif *License*: unknown *Contributors*: US Government employee

File:Claude stout brinegar.jpg *Source*: http://en.wikipedia.org/w/index.php?title=File:Claude_stout_brinegar.jpg *License*: unknown *Contributors*: ABrocke, Liesel

File:William thaddeus coleman.jpg *Source*: http://en.wikipedia.org/w/index.php?title=File:William_thaddeus_coleman.jpg *License*: unknown *Contributors*: Liesel

File:Brock Adams.jpg *Source*: http://en.wikipedia.org/w/index.php?title=File:Brock_Adams.jpg *License*: unknown *Contributors*: US Gov

File:Neil Goldschmidt.jpg *Source*: http://en.wikipedia.org/w/index.php?title=File:Neil_Goldschmidt.jpg *License*: unknown *Contributors*: Original uploader was Aboutmovies at en.wikipedia

Image:Drew lewis.jpg *Source*: http://en.wikipedia.org/w/index.php?title=File:Drew_lewis.jpg *License*: unknown *Contributors*: US Government

File:Elizabeth Dole official photo.jpg *Source*: http://en.wikipedia.org/w/index.php?title=File:Elizabeth_Dole_official_photo.jpg *License*: unknown *Contributors*: United States Congress

Image:James burnley IV.jpg *Source*: http://en.wikipedia.org/w/index.php?title=File:James_burnley_IV.jpg *License*: unknown *Contributors*: Liesel

File:Samuel Knox Skinner.jpg *Source*: http://en.wikipedia.org/w/index.php?title=File:Samuel_Knox_Skinner.jpg *License*: unknown *Contributors*: U.S. Department of Defense

File:Andrew Card award crop.jpg *Source*: http://en.wikipedia.org/w/index.php?title=File:Andrew_Card_award_crop.jpg *License*: unknown *Contributors*: U.S. Coast Guard Digital, PA3 Amy Thomas

File:Federico pena.jpg *Source*: http://en.wikipedia.org/w/index.php?title=File:Federico_pena.jpg *License*: unknown *Contributors*: Docu, Lukas skywalker

File:Slater rodney.jpg *Source*: http://en.wikipedia.org/w/index.php?title=File:Slater_rodney.jpg *License*: unknown *Contributors*: Infrogmation, Lukas skywalker, Materialscientist

File:Norman Mineta, official portrait, DOT.jpg *Source*: http://en.wikipedia.org/w/index.php?title=File:Norman_Mineta,_official_portrait,_DOT.jpg *License*: unknown *Contributors*: Superm401, Tom

File:Mary Peters official DOT portrait.jpg *Source*: http://en.wikipedia.org/w/index.php?title=File:Mary_Peters_official_DOT_portrait.jpg *License*: unknown *Contributors*: Department of Transportation

File:Ray LaHood Habitat for Humanity 2009 131156.jpg *Source*: http://en.wikipedia.org/w/index.php?title=File:Ray_LaHood_Habitat_for_Humanity_2009_131156.jpg *License*: unknown *Contributors*: Closeapple, SchuminWeb, Tony Wills

Image:Map of federal states.svg *Source*: http://en.wikipedia.org/w/index.php?title=File:Map_of_federal_states.svg *License*: unknown *Contributors*: User:Lokal_Profil

Image:Map of USA with state names.svg *Source*: http://en.wikipedia.org/w/index.php?title=File:Map_of_USA_with_state_names.svg *License*: unknown *Contributors*: User:Andrew c

Image:Mapa Mexico 2010.PNG *Source*: http://en.wikipedia.org/w/index.php?title=File:Mapa_Mexico_2010.PNG *License*: unknown *Contributors*: User:Hpav7

Image:Carte suisse2.png *Source*: http://en.wikipedia.org/w/index.php?title=File:Carte_suisse2.png *License*: unknown *Contributors*: Mschlindwein, Tschubby, 1 anonymous edits

Image:Comunidades autónomas de España.svg *Source*: http://en.wikipedia.org/w/index.php?title=File:Comunidades_autónomas_de_España.svg *License*: unknown *Contributors*: Rodriguillo

Image:CentralAmerica1860MapSmall.jpg *Source*: http://en.wikipedia.org/w/index.php?title=File:CentralAmerica1860MapSmall.jpg *License*: unknown *Contributors*: Unknown, irrelevant due to age. (Book located, scanned, and image uploaded to Wikipedia by Infrogmation at en.wikipedia, which of course has absolutely nothing whatsoever to do with authorship)

File:Flag of Argentina.svg *Source*: http://en.wikipedia.org/w/index.php?title=File:Flag_of_Argentina.svg *License*: unknown *Contributors*: User:Dbenbenn

File:Flag of Australia.svg *Source*: http://en.wikipedia.org/w/index.php?title=File:Flag_of_Australia.svg *License*: unknown *Contributors*: Anomie, Mifter

File:Flag of Austria.svg *Source*: http://en.wikipedia.org/w/index.php?title=File:Flag_of_Austria.svg *License*: unknown *Contributors*: User:SKopp

File:Flag of Belgium (civil).svg *Source*: http://en.wikipedia.org/w/index.php?title=File:Flag_of_Belgium_(civil).svg *License*: unknown *Contributors*: Bean49, David Descamps, Dbenbenn, Denelson83, Evanc0912, Fry1989, Gabriel trzy, Howcome, IvanOS, Ms2ger, Nightstallion, Oreo Priest, Rocket000, Rodejong, Sir Iain, ThomasPusch, Warddr, Zscout370, 4 anonymous edits

File:Flag of Bosnia and Herzegovina.svg *Source*: http://en.wikipedia.org/w/index.php?title=File:Flag_of_Bosnia_and_Herzegovina.svg *License*: unknown *Contributors*: User:Kseferovic

File:Flag of Brazil.svg *Source*: http://en.wikipedia.org/w/index.php?title=File:Flag_of_Brazil.svg *License*: unknown *Contributors*: Anomie

File:Flag of Canada.svg *Source*: http://en.wikipedia.org/w/index.php?title=File:Flag_of_Canada.svg *License*: unknown *Contributors*: Anomie

File:Flag of the Comoros.svg *Source*: http://en.wikipedia.org/w/index.php?title=File:Flag_of_the_Comoros.svg *License*: unknown *Contributors*: User:SKopp

File:Flag of Ethiopia.svg *Source*: http://en.wikipedia.org/w/index.php?title=File:Flag_of_Ethiopia.svg *License*: unknown *Contributors*: Aaker, Anime Addict AA, Antemister, Cycn, F l a n k e r, Fry1989, GoodMorningEthiopia, Happenstance, Homo lupus, Huhsunqu, Ixfd64, Klemen Kocjancic, MartinThoma, Mattes, Mozzan, Neq00, OAlexander, Pumbaa80, Rainforest tropicana, Reisio, SKopp, Smooth O, Spiritia, ThomasPusch, Torstein, Wsiegmund, Zscout370, 16 anonymous edits

File:Flag of Germany.svg *Source*: http://en.wikipedia.org/w/index.php?title=File:Flag_of_Germany.svg *License*: unknown *Contributors*: Anomie

File:Flag of India.svg *Source*: http://en.wikipedia.org/w/index.php?title=File:Flag_of_India.svg *License*: unknown *Contributors*: Anomie, Mifter

File:Flag of Iraq.svg *Source*: http://en.wikipedia.org/w/index.php?title=File:Flag_of_Iraq.svg *License*: unknown *Contributors*: User:Hoshie, User:Militaryace

File:Flag of Malaysia.svg *Source*: http://en.wikipedia.org/w/index.php?title=File:Flag_of_Malaysia.svg *License*: unknown *Contributors*: User:SKopp

File:Flag of Mexico.svg *Source*: http://en.wikipedia.org/w/index.php?title=File:Flag_of_Mexico.svg *License*: unknown *Contributors*: User:AlexCovarrubias

File:Flag of Federated States of Micronesia.svg *Source*: http://en.wikipedia.org/w/index.php?title=File:Flag_of_Federated_States_of_Micronesia.svg *License*: unknown *Contributors*: -

File:Flag of Nepal.svg *Source*: http://en.wikipedia.org/w/index.php?title=File:Flag_of_Nepal.svg *License*: unknown *Contributors*: User:Achim1999, User:Pumbaa80

File:Flag of Nigeria.svg *Source*: http://en.wikipedia.org/w/index.php?title=File:Flag_of_Nigeria.svg *License*: unknown *Contributors*: User:Jhs

File:Flag of Pakistan.svg *Source*: http://en.wikipedia.org/w/index.php?title=File:Flag_of_Pakistan.svg *License*: unknown *Contributors*: User:Zscout370

File:Flag of Russia.svg *Source*: http://en.wikipedia.org/w/index.php?title=File:Flag_of_Russia.svg *License*: unknown *Contributors*: Anomie

File:Flag of Saint Kitts and Nevis.svg *Source*: http://en.wikipedia.org/w/index.php?title=File:Flag_of_Saint_Kitts_and_Nevis.svg *License*: unknown *Contributors*: User:Pumbaa80

File:Flag of South Africa.svg *Source*: http://en.wikipedia.org/w/index.php?title=File:Flag_of_South_Africa.svg *License*: unknown *Contributors*: Adriaan, Anime Addict AA, AnonMoos, BRUTE, Daemonic Kangaroo, Dnik, Duduziq, Dzordzm, Fry1989, Homo lupus, Jappalang, Juliancolton, Kam Solusar, Klemen Kocjancic, Klymene, Lexxyy, Mahahahaneapneap, Manuelt15, Moviedefender, NeverDoING, Ninane, Poznaniak, Przemub, SKopp, ThePCKid, ThomasPusch, Tvdm, Ultratomio, Vzb83, Zscout370, 35 anonymous edits

File:Flag of South Sudan.svg *Source*: http://en.wikipedia.org/w/index.php?title=File:Flag_of_South_Sudan.svg *License*: unknown *Contributors*: User:Achim1999

File:Flag of Spain.svg *Source*: http://en.wikipedia.org/w/index.php?title=File:Flag_of_Spain.svg *License*: unknown *Contributors*: Anomie

File:Flag of Sudan.svg *Source*: http://en.wikipedia.org/w/index.php?title=File:Flag_of_Sudan.svg *License*: unknown *Contributors*: Vzb83

File:Flag of Switzerland.svg *Source*: http://en.wikipedia.org/w/index.php?title=File:Flag_of_Switzerland.svg *License*: unknown *Contributors*: User:-xfi-, User:Marc Mongenet, User:Zscout370

File:Flag of the United Arab Emirates.svg *Source*: http://en.wikipedia.org/w/index.php?title=File:Flag_of_the_United_Arab_Emirates.svg *License*: unknown *Contributors*: Anime Addict AA, Avala, Dbenbenn, Duduziq, F l a n k e r, Fry1989, Fukaumi, Gryffindor, Guanaco, Homo lupus, Kacir, Klemen Kocjancic, Krun, Madden, Neq00, Nightstallion, Piccadilly Circus, Pmsyyz, RamzyAbueita, Zscout370, 4 anonymous edits

File:Flag of the United States.svg *Source*: http://en.wikipedia.org/w/index.php?title=File:Flag_of_the_United_States.svg *License*: unknown *Contributors*: Anomie

File:Flag of Venezuela.svg *Source*: http://en.wikipedia.org/w/index.php?title=File:Flag_of_Venezuela.svg *License*: unknown *Contributors*: Alkari, Bastique, Denelson83, DerFussi, Fry1989, George McFinnigan, Herbythyme, Homo lupus, Huhsunqu, Infrogmation, K21edgo, Klemen Kocjancic, Ludger1961, Neq00, Nightstallion, Reisio, Rupert Pupkin, Sarang, ThomasPusch, Vzb83, Wikisole, Zscout370, 12 anonymous edits

file:Fairfax County Public Schools Logo.svg *Source*: http://en.wikipedia.org/w/index.php?title=File:Fairfax_County_Public_Schools_Logo.svg *License*: unknown *Contributors*: Connormah, Tbhotch

File:Fairfax High School.jpg *Source*: http://en.wikipedia.org/w/index.php?title=File:Fairfax_High_School.jpg *License*: unknown *Contributors*: Kawah Wahidi

MIX
Papier aus verantwortungsvollen Quellen
Paper from responsible sources
FSC® C105338

Printed by Books on Demand GmbH, Norderstedt / Germany